Advanced Pr
Longevity Gu

"Peter's passion for unlocking the potential of human health is boundless, and he's made it his mission to share that knowledge with the world. Peter and I initially joined forces in founding Fountain Life and then in writing our #1 *NYTimes* bestseller *Life Force*. His latest work, *Longevity Guidebook*, is packed with the latest, cutting-edge science and real-world strategies that will not only extend your healthspan but supercharge your energy, focus, and vitality. Peter has taken everything we learned and personalized it into an actionable, practical guide that anyone can use to take massive action right now. If you're ready to transform your body and take control of the future of your health, *Longevity Guidebook* is your blueprint for success."

Tony Robbins, America's #1 Life & Business Strategist, Philanthropist, Author #1 *New York Times* Bestseller

"Peter has taken the idea of longevity escape velocity and transformed it into a roadmap we can all follow. His commitment goes far beyond personal ambition; he is carving a path for humanity's future by harnessing the latest health technology and biotechnology. As we move toward breakthroughs in age-reversal therapies and await the monumental achievement of the $101 Million XPRIZE Healthspan, Peter's *Longevity Guidebook* offers practical strategies that empower us today. It's a guide for those who want to take control of their health now while anticipating the extraordinary advancements yet to come."

Ray Kurzweil, Inventor, Author, Futurist, Co-Founder, Singularity University, Principal Researcher and AI Visionary at Google

"Peter's *Longevity Guidebook* is a must-read for anyone seeking maximal health and performance. From advanced diagnostics and therapeutics to simple but important lifestyle changes, Peter outlines what is possible today and what is coming in the decade ahead."

David Sinclair, PhD, Geneticist and Professor at Harvard Medical School

"Peter Diamandis is uniquely suited to look at the future of health and longevity through the lens of an engineer and a truly prescient futurist. His work with the $101 million XPRIZE Healthspan and Fountain Life has the potential to surface and test the next generation of therapeutics. I'm excited to continue to follow his journey."

Rhonda Patrick, PhD, Biochemist, Host, FoundMyFitness podcast

"We are in the midst of a healthspan revolution that has the potential to cure and prevent most diseases—adding decades of youth for each of us. Peter and I share a vision of a future where such health may eventually be universally affordable and uncapped. *Longevity Guidebook* masterfully explores how advancements in diagnostics and therapeutics, such as epigenetic reprogramming and regenerative medicine, can drive us toward this future. This book offers a clear step-by-step methodology for shaping lifestyle to bridge us to this future and lays out a path to a world where longevity isn't just extended—it's reimagined."

George Church, PhD,
Professor at Harvard Medical School and MIT and Author of *Regenesis*

"*Longevity Guidebook* offers not just the 'what,' but the 'how' of enhancing your health. What awaits you here? Simply put, it is a masterclass in actionable insights grounded in rigorous science, yet accessible and eminently practical. Peter lays out a clear, evidence-based blueprint designed to improve your well-being and longevity. For me, this wasn't just a book I read; it was a formative guide to a healthier, longer life."

Matthew Walker, PhD, Professor of Neuroscience and Psychology,
University of California, Berkeley; Author of *Why We Sleep*

"In *Longevity Guidebook,* Peter Diamandis brilliantly integrates cutting-edge science with functional medicine principles to give readers a clear and practical roadmap to achieve optimal health and longevity. He also offers a clear vision of the therapeutic breakthroughs racing toward us that will add decades to our healthspan. This book is a must-read and powerful resource for anyone looking to live longer and better by addressing the root causes of aging."

Mark Hyman, MD, Functional Medicine Practitioner,
New York Times Bestselling Author

"Diamandis delivers a practical guidebook of clear and actionable guidance on what you can do today to extend your healthspan. This book will help you make better decisions every day on what you eat, how much you sleep, your exercise plan, and your mindset. The *Longevity Guidebook* is a foundational read for anyone interested in longevity.

> **Eric Verdin**, MD, CEO, The Buck Institute; Associate Professor, University of California, San Francisco School of Medicine; Fellow of the American Association for the Advancement of Science

"Peter's commitment to understanding and advocating for women's health has been extraordinary, supporting our work from its inception. *Longevity Guidebook* helps shine a spotlight on the often-overlooked connection between ovarian health and overall longevity and the importance of ovaries as the canary in the coal mine for aging. This book raises awareness about the unique challenges women face."

> **Jennifer Garrison**, PhD, Co-founder and Executive Director, ProductiveHealth.org; Center for Healthy Aging in Women, The Buck Institute

"Peter Diamandis shares a compelling how-to guidebook on the latest diagnostic and therapeutic tools we can all access to extend our healthspan. But what truly stands out in his book, *Longevity Guidebook,* is how much we can achieve by simply adjusting our lifestyle with consistency, creating significant, lasting change."

> **Deepak Chopra**, MD, Alternative Medicine Advocate, *New York Times* Bestselling Author, and Founder, Chopra Foundation

"Peter Diamandis is the Pope of Hope—one of the most positive-minded and action-oriented leaders I know. His current mission is to help make all of us live healthier, longer lives. This guidebook makes longevity accessible, understandable, and actionable. It's not science fiction; it's science fact, and it's available now."

> **Dean Kamen**, Inventor, Entrepreneur, Founder of FIRST Inspires, ARMI, and DEKA R&D

"In *Longevity Guidebook*, Moonshot XPRIZE pioneer Peter Diamandis offers far more than a manual on extending your lifespan. It's a guide to becoming the CEO of your own well-being and leading a healthier, more productive life."

Marc Benioff, Chair and CEO, Salesforce

"I've been friends with Peter Diamandis for decades, and his visionary work is always at least ten years ahead of everyone else. In his latest book, *Longevity Guidebook*, Dr. Diamandis does a superb job providing a clear step-by-step science-based methodology for shaping a lifestyle to elevate both the quality and quantity of your life. Highly recommended!"

Dean Ornish, MD, Renowned Physician, Researcher, and #1 *New York Times* Bestselling Author

"Peter Diamandis is a warrior for our health and is tireless in his efforts. In the same way he led the birth of the commercial spaceflight industry twenty years ago, he has turned his considerable attention, capital, and relationships toward unlocking extended healthspan for everyone. *Longevity Guidebook* is a clear and compelling blueprint, a tour-de-force, detailing what each of us can do today to unlock the coming biomedical breakthroughs of tomorrow."

Jamie Justice, PhD, EVP, Health, XPRIZE Foundation, Executive Director, XPRIZE Healthspan

"In *Longevity Guidebook*, Peter Diamandis expertly bridges the gap between cutting-edge science and everyday health practices. As a surgeon and advocate for lifestyle medicine, I am impressed with how Peter demystifies complex health principles and presents them in an accessible, actionable way. This book is a roadmap for optimizing your body and mind for the long haul. I urge you to read it, follow it, and keep it within easy reach."

Mehmet Oz, MD, Professor Emeritus, Columbia University

"*Longevity Guidebook* is more than just a manual for health; it's a gateway into the future of biohacking. Peter stacks his rocket scientist medical doctor brain with insights from the top research scientists and sprinkles in wisdom from experts, creating the future of diagnostics, supplementation, and regenerative medicine. This book equips you with the tools to take control of your biology and push the boundaries of what's possible for your body and mind. Don't just live a healthy life. Live way longer *and* better than you can imagine."

Dave Asprey, Founder of Bulletproof & Upgrade Labs, Father of Biohacking

"Peter Diamandis absolutely nailed it with the *Longevity Guidebook*, merging cutting-edge science with real, actionable advice. This isn't just a guide; it's a game plan for upgrading your daily habits and tapping into the most advanced diagnostics and therapies available today. If you're ready to become the CEO of your health and push the limits of human potential and longevity, this is the book you've been waiting for."

Gary Brecka, Human Biologist, Founder of The Ultimate Human, and Co-Founder of 10X Health

LONGEVITY GUIDEBOOK

How to Slow, Stop, and Reverse Aging — and NOT Die From Something Stupid

Other Books by Peter

Abundance: The Future Is Better Than You Think (2012)

Bold: How to Go Big, Create Wealth, and Impact the World (2015)

The Future Is Faster Than You Think: How Converging Technologies Are Transforming Business, Industries, and Our Lives (2020)

Life Force: How New Breakthroughs in Precision Medicine Can Transform the Quality of Your Life and Those You Love (2022)

ExO 2.0: The Playbook for Transforming Your Organization (2023)

LONGEVITY GUIDEBOOK

*How to Slow, Stop, and Reverse Aging
— and NOT Die From Something Stupid*

Peter H. Diamandis, MD

**With a chapter on women's health
by Helen Messier, PhD, MD**

ethos
collective

LONGEVITY GUIDEBOOK © 2025 by Peter Diamandis. All rights reserved.

Printed in the United States of America

Published by Igniting Souls
PO Box 43, Powell, OH 43065
IgnitingSouls.com

This book contains material protected under international and federal copyright laws and treaties. Any unauthorized reprint or use of this material is prohibited. No part of this book may be reproduced or transmitted in any form or by any means, electronic or mechanical, including photocopying, recording, or by any information storage and retrieval system, without express written permission from the author.

LCCN: 2024924395
Paperback ISBN: 978-1-63680-427-9
Hardcover ISBN: 978-1-63680-428-6
e-book ISBN: 978-1-63680-429-3

Available in paperback, hardcover, e-book, and audiobook.

Some names and identifying details may have been changed to protect the privacy of individuals.

DEDICATION

To my Abundance360, Abundance Platinum, Fountain Life,
XPRIZE, and Singularity University communities.
Thank you for giving me the gift of allowing me
to share my passions and learnings with you.

All Book Profits Will Be Contributed to XPRIZE

Please note that all profits from this book and the audiobook
are being donated to the XPRIZE Foundation to support
the work they are doing with the XPRIZE Healthspan
to reverse the ravages of aging by ten to twenty years.

"Life is short... Until you extend it!"
—*Peter H. Diamandis, MD*

Scan the QR code for the
latest updates, addendums,
and resources.

LongevityGuidebook.com/resources

Contents

Why I Wrote This Book		xvi
Introduction		xviii
Peter's Disclosures		xxvi
Chapter 1	Fueling Your Future: The Longevity Diet	1
Chapter 2	Exercise Blueprint: Build Strength, Endurance, & Longevity	27
Chapter 3	Mastering Sleep: Simple Practices That Work	39
Chapter 4	Don't Die from Something Stupid: Breakthroughs That Can Save Your Life	53
Chapter 5	Your Longevity Pharmacy: Medications, Supplements, & Cutting-Edge Therapies	69
Chapter 6	Longevity Mindset & Happiness Hormones	120
Chapter 7	Winning Habits: Routines That Keep You Young	131
Chapter 8	Women's Health: Longevity Through Every Stage of Life	144

APPENDICES

Appendix A: Blue Zone Wisdom	182
Appendix B: Abundance360	186
Appendix C: Join Peter for His Annual Longevity Platinum Trip – "A Virtual Blue Zone"	190
Additional Tools & Resources	194
Endnotes	196

Why I Wrote This Book

WHY I WROTE THIS BOOK

Today, at the age of 63, I find myself in peak health—physically, mentally, and energetically. A wide range of performance metrics and biomarker data bolster this belief. I've gotten here not by luck but by effort and prioritizing healthspan. I'm on a personal mission to maintain this optimal health for the next decade so I can intercept the next generation of therapeutics under development that are promising to slow, stop, and even reverse aging.

Over the past decade, my primary focus has been to study the field of human longevity (more specifically, healthspan). I've immersed myself in research and devoured publications in biotechnology, nutrition, exercise strategies, sleep, and, most recently, artificial intelligence in the service of health and longevity. I've absorbed information from a myriad of podcasts, interviewed top scientists on my podcast *Moonshots,* and engaged in countless on-stage interviews during my Abundance Summit and Longevity Platinum Trips. Finally, I'm proud to hold in close company dozens of top scientists and physicians who work closely with me in the various companies I've founded or co-founded, including Fountain Life, Life Force, Celularity, Vaxxinity, Abundance360, Singularity University, and the XPRIZE Foundation.

The aim of the *Longevity Guidebook* is twofold: to simplify and prescribe. While there is a ton of information out there on longevity, it's hard to keep it all straight and make it practical and usable. That's where this book comes in. I've designed this guidebook to be digestible within a few hours and to serve as a practical reference guide that you can easily turn to as needed. I hope this book will serve as a bridge to the next wave of healthspan-extending breakthroughs expected over the decade ahead.

Don't underestimate the value of this book due to its brevity. The challenge was not to accumulate material but to curate it to be easily consumable while maintaining rigorous scientific research backing. After all, what's the point of amassing a library of health books if they're left unread or if you don't devise a clear, actionable plan for your health and well-being?

I sincerely hope you find this book beneficial, but more importantly, I hope it inspires you to prioritize your health. In today's world, health is the new wealth. To quote one of my favorite sayings, "The individual who has health has a thousand dreams. The one without it has but one." Here's to an extraordinary decade ahead.

Peter H. Diamandis, MD
Founder, Executive Chairman, XPRIZE Foundation
Executive Founder, Singularity University, and
Founder, Abundance360 & Abundance Platinum
Co-Founder and Executive Chairman, Fountain Life

Introduction

We are in the midst of a healthspan revolution.

In this book, I'm sharing my longevity practices—what I'm doing to extend my healthspan. I've synthesized these learnings and protocols from over three hundred interviews during my Longevity Platinum Trips, *Moonshots* podcasts, and in consultation with the medical team at Fountain Life.

In the first part of this book, we'll look briefly at the context in which this healthspan revolution is taking place and explore the concept of "longevity escape velocity (LEV)." Simply put, converging exponential technologies such as AI, genomics, and other "omics technologies," CRISPR, gene therapy, cellular medicine, and invasive and non-invasive sensors allow us to understand why we age and how to slow, stop, and perhaps even *reverse* aging.

Experts predict that breakthroughs within the next ten years *will enable us to add decades to our healthspan.*

Healthspan refers to the period of life spent in good health, *free from chronic diseases and the disabilities of aging.* It emphasizes the quality of life lived rather than just the duration. In contrast, lifespan is the total duration of an individual's life, regardless of their health condition. While lifespan denotes the years lived, healthspan describes the years lived actively and in good health.

Your mission is, therefore, to maximize your healthspan and vitality by using the most advanced diagnostics to catch diseases early, enabling you to stay healthy long enough to intercept the additional breakthroughs racing toward us.

Approaching Longevity Escape Velocity

So, how long might we all live?

When I was in medical school in the late 1980s, I distinctly remember one Sunday afternoon watching a documentary on the topic of "long-lived sea life."

It turns out that bowhead whales from the Arctic can live for 200 years, and Greenland sharks double that life expectancy with an impressive lifespan of 400 and perhaps even 500 years.[1] These sharks can even have pups (babies) at 200 years old.

INTRODUCTION

I remember thinking, *If they can live that long, why can't we?*

As an engineer, I figured it was either a hardware or software problem, something we would eventually fix by altering our genome or repairing our cells, and that it wouldn't be too long before we could solve it.

My dear friend Ray Kurzweil, co-founder of Singularity University and futurist at Google, and Aubrey de Grey, a leading biomedical gerontologist, both speak about a concept called *longevity escape velocity*.[2] It's an intriguing notion that goes something like this:

Today, by some estimates, science and medicine are adding about three months to your lifespan every year. In the near future, additional scientific breakthroughs will extend your lifespan *by more than a year* for every year you remain alive. Once that happens, we can begin thinking about true longevity.

That concept is called "longevity escape velocity."

I've asked many of the leading scientists in the field of health and longevity for their predictions on when we might expect to reach LEV. The answers vary widely, but Kurzweil and de Grey are two leaders in the field who give me the greatest confidence.

Kurzweil's written works have become intellectual landmarks. His *New York Times* bestsellers *The Singularity Is Near* (2005) and his most recent book, *The Singularity is Nearer* (2024), are definitive texts in the field of artificial intelligence and exponential technologies.

Beyond his brilliance, Kurzweil is most famous for his accurate predictions about the future. To date, his written predictions—147 in total—have demonstrated an astounding 86 percent accuracy (see his Wikipedia entry for more details on his predictions).

When asked for his prediction about when we'll reach longevity escape velocity, he said, "For those in reasonably good shape and with reasonable means, *I believe they will have access to longevity escape velocity by the end of 2030.*"

Dr. George Church, PhD, of Harvard Medical School echoes a similar timeframe. A towering figure at the forefront of genetics, synthetic biology, and longevity research, Church's pioneering work transcends the conventional boundaries of science, having revolutionized genome sequencing, gene editing, gene synthesis, and the burgeoning field of age-related studies. He has founded over fifty cutting-edge synthetic biology companies, ranging from companies that are regrowing human organs to one (Colossal) that is bringing back extinct animals.

According to Dr. Church, "The exponential technologies that have improved the speed and cost of reading, writing, and editing DNA and of implementing gene therapies now apply to the category of age reversal."

He adds, "Age-reversal advances could mean that we reach longevity escape velocity in a decade or two, within the range of the next one or two clinical trials."

So, what does that mean? Can we extend the healthy human lifespan past today's record of 122? Can humans live past 200 years old—or even indefinitely?

Ray Kurzweil & Peter at Founding Conference of Singularity University

I recently sat down with a friend, Dr. David Sinclair, Professor of Genetics at Harvard Medical School and author of *Lifespan*, to discuss these topics during an episode of my *Moonshots* podcast.

I opened with a question about how long humans might be able to live: "Is there an upper age limit?"

Sinclair's answer was inspiring: "There is no biological limit... of course, there isn't," he began. "We are the same stuff as a whale that can live a lot longer than us (200 years); we're built of the same stuff as a tortoise, pretty much the same stuff as trees that can live thousands of years."

"I'm putting my career on the line," he continued. "It's a software problem, and what's interesting about biology is that software encodes the ability to rebuild the hardware. So, we need to reset the software. And when we do that in my lab, we find that tissues regenerate in animals; organoids, which are mini human organs, regenerate. They fix themselves, and they function like they are new again. So it is, in my view, 99 percent a software problem."

How Long Will You Live? Heredity vs. Lifestyle

How long you live is a function of many factors: where you were born, your genetics, your diet, your lifestyle, and your mindset. Most people imagine that longevity is mostly inherited and that the genetic cards you've been dealt will predetermine your lifespan.

You may be surprised by the truth.

INTRODUCTION

In 2018, after the analysis of a 54-million-person ancestry database, scientists announced that lifespan has little to do with genes.[3]

In fact, heritability is accountable for roughly 7 percent of your longevity.[4]

The highest estimates for the impact of heredity on longevity are around 30 percent—which still means, at a minimum, you're 70 percent in control of how you age.

The power to shape your healthspan is much more in your hands than you might have imagined.

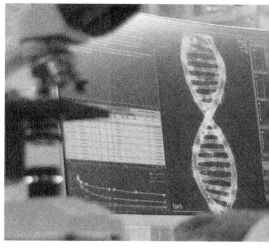

Heredity has a limited impact on your longevity compared to lifestyle choices

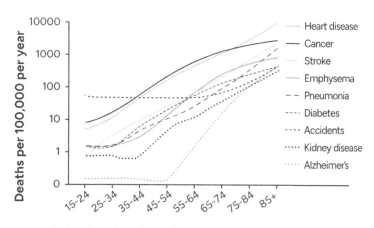

Today, deaths from chronic disease grow exponentially with age.

There is no question that aging is the number one risk factor that correlates with chronic disease, and a goal of slowing, stopping, and potentially reversing aging and, therefore, disease should be at the top of everyone's health objectives.

As we will see in the chapters ahead, advances in diagnostics and therapeutics enable us to find diseases at inception and cure or reverse their course. What was once outside the power of medicine is now very much within reach.

Our Goal: Extending Healthspan

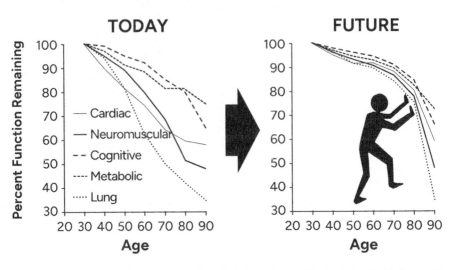

Today, we lose function with age, a slow and steady decline after age 30. Age is the #1 risk factor with chronic disease. Our future goal must be to intervene, slow (or stop) aging and prevent functional loss, to maintain vitality.

The Journey Ahead

In the chapters that follow, I'll share details about what I've learned, synthesized, and adopted. *This is the practical guide on what you might consider doing to become the CEO of your own health.* My experience is that many longevity, health, and diet "how-to" books are way too long, include too much theory, and are impractical to follow. That's why I have attempted to create something that is simple and easy to consume, reference, and implement. May

> "I implore you to consider adopting some of these protocols and strategies and to think through what you would do with an extra thirty years of healthy life."
>
> —Peter H. Diamandis, MD

Some Words of Inspiration & Wisdom

"I believe that aging is a disease. I believe it is treatable. I believe we can treat it within our lifetimes. And in doing so, I believe everything we know about human health will be fundamentally changed."
—David Sinclair, PhD

"Our aim should be to help our patients die young as late as possible."
—Tenley Albright, MD

"It's likely that we're just ten years away from the point that the general public will hit longevity escape velocity."
—Ray Kurzweil

"The best way to enjoy your aging process is to take care of your body and mind, to keep learning and growing, and to find joy and purpose in the present moment."
—Deepak Chopra, MD

"It is health that is the real wealth, and not pieces of gold and silver."
—Mahatma Gandhi

"Knowledge is the antidote to fear."
—Ralph Waldo Emerson

"Aging is malleable. We can manipulate it to live healthier, longer lives."
—Nir Barzilai, MD

"The first person to live to 150 has already been born."
—Aubrey de Grey

"Longevity isn't just about adding years to your life. It's about adding life to your years."
—Dan Buettner

"Extending the healthy human lifespan will increase global abundance and uplift humanity. The equation is simple: longer, healthier lives mean more time spent at our productive best, which means more innovation."
—Peter H. Diamandis, MD

Peter's Disclosures

I am an educator, entrepreneur, and scientist. While I completed my medical degree at Harvard Medical School, *I am not a clinician and cannot make clinical recommendations for the prevention or treatment of any disease. In making the suggestions outlined in this book, I am expressing what I have learned through research and interviews and sharing what I am personally doing for my health.*

No one should start taking any supplements or medications without first checking with their physician. Some supplements can be dangerous for people with certain pre-existing medical conditions, and supplements can interfere with some prescription medicines. Supplements can and will affect people differently.

Regarding supplements, the FDA is limited to post-market enforcement because, unlike prescription drugs that must be proven safe and effective for their intended use before marketing, there are no provisions in the law for the FDA to approve dietary supplements for safety before they reach the consumer.

Note: The evidence of benefit for most supplements comes from laboratory experiments and/or from epidemiology data—not from human clinical trials.

Supplements should only be purchased from trusted retailers and brands. Testing has shown that many supplements are tainted with unlisted ingredients and/or do not contain the amount of the supplement listed on their label.

For more detailed recommendations for sleep, exercise, diet, and supplements, you can also turn to a book called *Life Force*, which I co-authored with Tony Robbins and Robert Hariri, MD, PhD. A bibliography of additional books I recommend on these topics is available here: LongevityGuidebook.com/resources

I also want to disclose that I am a founder, co-founder, investor, and/or advisor to a number of the companies discussed in this book. That shouldn't come as a surprise since I would naturally want to invest in and/or be involved in those companies and technologies I believe make the biggest dent in this field. I will do my best to disclose my relationship to those with whom I have a financial relationship and, where possible, offer alternative companies that deliver similar products or services.

Chapter 1
Fueling Your Future: The Longevity Diet

"Let your food be your medicine, and your medicine be your food."
—**Hippocrates**

You very literally are what you eat. The nutrients (or non-nutrients) we consume become our bodies and our minds.

What you eat and drink, how much you eat, when you eat, the way you eat, and even the order in which you eat foods can be critically important.

No singular diet serves everyone the best. Thousands of diet books have been published, with over five million sold in the US alone. So, who do you believe? Where do you start?

My hope is that the details outlined in this chapter will be fundamental to everyone. But ultimately, what is best for you will depend upon your genetics, age, microbiome, environmental exposures, health objectives, and your physician's advice.

After interviewing dozens of scientists and physicians in the field, researching the topic, and experimenting on myself, below is a summary of my diet practices. Beyond what I do, I want to explain why I made these choices so you can decide what habits and practices to adopt. Without question, there are small and important steps you can take right now to increase your potential healthspan and ensure your diet works for you.

Peter and his PHD Ventures team.

My Longevity Diet

What I Do NOT Eat

No Added Sugar or High Glycemic Foods: For most of human history, sugar was a rare and precious resource, primarily found in fruits and honey, which were consumed seasonally and in moderation. It wasn't until the advent of agriculture and, later, the Industrial Revolution that refined sugar became widely available. As a result, our genetics and physiology, which evolved over millions of years, are not equipped to handle the excessive and constant intake of sugar and high-glycemic foods that characterize today's diet.

The chronic overconsumption of sugar has been linked to a slew of serious health issues, including obesity, type 2 diabetes, heart disease, cancer, chronic inflammation, and neurodegenerative disease.

As brutal as it might sound, I think of refined sugar as a poison. I try to avoid sugar, simple carbohydrates, and processed starches. The human body never evolved to consume the levels of sugar in most diets today. Human physiology evolved on a diet containing very little added sugar and virtually no refined carbohydrates. In fact, refined sugar probably entered our diets by accident. Sugarcane was likely primarily a "fodder" crop used to fatten pigs, though humans may have chewed on the stalks from time to time.

Sugar is a poison.

The effects of added sugar intake can be devastating, including higher blood pressure, cardiac- and neuro-inflammation, weight gain, diabetes, fatty liver disease, and providing fuel for cancer.[5] Increased blood sugar levels are directly linked to an increased risk of heart attack and stroke. On the other hand, reduced blood sugar levels have been linked to lower blood pressure and lower cholesterol and have been shown to reduce the risk of heart attack, stroke, and heart-related death. I believe many of today's children's cereals that are overly laden with sugar should carry the FDA's black-box health warnings.

The average American consumes about 154 pounds of sugar annually. So, if it's so bad for us, why do so many people knowingly consume so much? Sugar is damn addictive, with some scientific research suggesting it can be as addictive as drugs like cocaine. A 2007 study published in *PLOS ONE* (*Public Library of Science*, peer-reviewed open access mega journal) found that when given a choice, rats consistently preferred sugar water over cocaine, even when they were already addicted to cocaine.[6]

This preference was observed in 94 percent of the rats tested. The addictive power of sugar lies in its ability to trigger the release of dopamine in the brain's reward center, very much like addictive drugs. According to a 2013 study in the *American Journal of Clinical Nutrition*, high-sugar milkshakes activated the same brain regions as cocaine in human subjects.[7] Moreover, a 2015 review in the *British Journal of Sports Medicine* concluded that sugar meets the criteria for a substance of abuse and may be addictive to some individuals.[8]

So, how do you kick such a habit? Every year, as part of Abundance Summit, I'm joined by nutritionist Dr. Guillermo Rodriguez Navarrete (Fellow of the American College of Nutrition and a member of the American Society for Nutrition) to take a segment of the Abundance membership through a "22 Day No-Sugar Challenge." We do this as a group on WhatsApp. The communal nature of "doing it together" is incredibly useful. It takes about three weeks for your brain and body to eliminate cravings for sweets and begin craving healthier foods that satisfy more of your actual nutritional needs. The good news is that you can break this addiction, and the results can be monumental.

In my discussions with Mark Hyman, MD, a physician and author of the bestselling book *Young Forever*, he noted: "When you eat sugar, it slows your metabolism down, and it increases your hunger hormones. So, you're hungrier, you're gaining weight, and you can't burn the fat." Sugar consumption increases levels of the hormone insulin, which has the primary purpose of turning that sugar into stored fat.

For males, as well as females, sugar intake results in significant hormonal problems—and perhaps most shockingly, new evidence proves that sugar shrinks your brain's hippocampus, which is your memory center, leading to poor memory and reduced overall brain volume.[9] So, the next time you have a sugar craving, think about how it may literally shrink your brain cells.

As Dr. Hyman remarks, "93 percent of Americans are metabolically unhealthy. They have high blood sugar, high cholesterol, high blood pressure; they're overweight, or they have already had a heart attack or a stroke." This means, in effect, that a mere 7 percent of us are in good metabolic health. And Dr. Hyman goes on to explain that it's "primarily driven by our diet."

Saying no to sugar can require tremendous willpower, combating the powerful signals from our brain's reward centers. And while willpower can be strong in the morning after a great night's sleep, it diminishes throughout the day for many, including myself. Over the years, I've adopted a few simple hacks to support my no-sugar diet objectives, including the following:

First, I don't bring any unhealthy desserts into the house, so they are never within easy reach. The best way to do this is to eat before you go food shopping; never go food shopping while you're hungry.

Second, I keep a number of healthy snacks within easy reach; these include walnuts and almonds. When I crave something sweet (and everyone does), on occasion, I will satisfy it with a bit of 75 percent dark chocolate or frozen blueberries.

Third, I've developed a reflex to say "no" to dessert whenever it's offered. Saying no immediately before the plate ever hits the table makes it easier than resisting a slice of cake sitting on the table in front of me throughout the dinner.

What Else I DON'T Eat (Dairy, Beef, Tuna)

Beyond avoiding sugar, I don't eat dairy products or beef. My food allergy testing indicates that I have an immune response to the casein protein component in dairy that can drive inflammation. I also avoid red meat, particularly beef, due to its high saturated fat content and the association of red meat with cancer—especially colon cancer—as well as cardiovascular disease. Although grass-fed beef can be a healthy part of some diets, it's easier for me to decrease consumption by avoiding it altogether.

While I *love* sushi, particularly maguro and toro, I've removed it from my diet for two reasons. The first reason is that large fish like tuna, swordfish, mackerel, and shark can contain very high levels of mercury. These large, long-lived fish, which consume other fish, concentrate mercury at the top of the food chain, and eating them can have a devastating impact on our cognitive health. The impact of mercury on brain function is crippling, causing impaired cognitive function, memory loss, and motor skills deterioration.

My dear friend and business partner, Tony Robbins, documented his battle with mercury poisoning in our book *Life Force*. Tony faced a significant health challenge when he discovered he was suffering from mercury poisoning. This issue came to light after Tony experienced a range of mysterious and severe health symptoms that included extreme fatigue, cognitive difficulties, and physical discomfort. The root cause of Tony's mercury poisoning was traced back to his diet, which included large amounts of fish, particularly swordfish and tuna.

For years, Tony had been eating these fish regularly, unaware that mercury was accumulating in his body. The situation reached a critical point when Tony underwent testing and found that his mercury levels were over 100 times higher than normal—a serious health crisis. Recognizing the gravity of the situation, Tony underwent a detoxification process to reduce the mercury levels in his body, involving years of chelation therapy, a treatment designed to remove heavy metals from the body.

There is a secondary reason for not eating these large fish, one focused on the health of our planet. Over the past fifty years, humanity has massively overfished our oceans and decimated 90 percent of the ocean's large fish populations.

There is one solution coming that might allow me to enjoy maguro and toro sushi guilt-free while addressing my dual concerns on mercury levels and environmental issues. That solution is called cultivated meats, also known as stem-cell-grown protein. Cultivated tuna, produced through advanced stem cell technology, begins by harvesting stem cells from a live tuna, which are then nurtured in a bioreactor containing a nutrient-rich medium that mimics the natural environment necessary for cell growth and differentiation. These stem cells proliferate and develop into muscle tissue, replicating the texture and flavor of traditional tuna without the need for extensive fishing. One of the key health benefits of cultivated tuna is the significant reduction in mercury and other toxic contaminants. Additionally, the process allows for the precise manipulation of the nutritional content, such as optimizing omega-3 fatty acid levels and reducing unhealthy fats, thereby enhancing the overall health benefits of the tuna.

This isn't just theory; the process is already commercially available and FDA-approved for cultivated chicken and will be soon for tuna as well. An Israeli startup called Wanda Fish has unveiled its first cultivated toro sashimi made from bluefin tuna, one of the most expensive fish in the world per pound.

Wanda Fish says their prototype dish meets the growing demand for bluefin tuna using a high-quality, pollution-free, and sustainable creation process. "A key focus in the creation of our product was achieving the same level of fat marbling as real bluefin toro sashimi to create the same look and mouthfeel," said Wanda Fish co-founder and CEO Daphna Heffetz. "Our bluefin tuna toro filet is sustainable and free of microplastics, mercury, and other chemical toxins all-too-commonly found in wild catch."

What I DO Eat

A Whole-Plant Diet

I proactively eat a whole-plant diet to the maximum extent possible. There's no question that consuming whole plants is a major health benefit. As such, I focus on spinach, broccoli, brussels sprouts, avocado, and asparagus, usually with a heavy helping of extra virgin olive oil. I'll typically have a Greek salad with added avocado and protein such as fish, chicken, or legumes for lunch.

As Dr. Helen Messier, chief medical officer of Fountain Life, says, when eating vegetables, "Eat the rainbow" (and that doesn't mean Skittles or Fruit Loops). Eating a wide variety of colored fruits, vegetables, and spices corresponds to the consumption of different phytochemicals, vitamins, minerals, and antioxidants, each with unique health benefits. For example, plants contain thousands of natural chemicals called "phytonutrients" or "phytochemicals," which are responsible for the color of the plants. These chemicals

FUELING YOUR FUTURE

have protective properties that can benefit human health when consumed. In addition, different colors often indicate different types of antioxidants, which are compounds that help protect our bodies from damage by free radicals, which can contribute to chronic disease and aging. When I see different colored vegetables on the table, I do my best to eat them all.

Vegetables, whole plants, are key to a longevity diet (the more colors the better).

Nuts, Beans, and Legumes for Protein

Another category of whole plants I actively select are those high in protein, such as nuts (typically macadamias, walnuts, and almonds), as well as properly soaked and prepared beans and legumes: soybeans, lentils, white beans, split peas, pinto beans, and black beans.

Pro-Tip on Snacking: Walnuts and Blueberries

Everyone snacks. At some point, you'll get hungry, and your willpower will drop. Now, the question is, what do you reach for to satisfy that hunger? What's within easy reach in your refrigerator or cupboard? Unfortunately, almost everyone, driven by the brain's reward centers, will reach for a sugary solution of cookies or ice cream. Once you give in, your blood sugar spikes, and insulin is released, creating a vicious cycle that can worsen your hunger.

I've solved the snacking problem in two ways, which I mentioned a few pages ago, but it is so important that it bears repeating. *First, I shop at the supermarket only after I've eaten, so I'm not tempted to purchase something I'll regret having access to when my willpower erodes. Second, I'll put healthy snacks within easy reach of my computer.* At this moment, I have a large open bag of walnuts a meter away from me and a bowl of frozen blueberries in the freezer. Both are rich in phytonutrients and antioxidants, which help combat oxidative

Nuts are a great source of protein.

stress and reduce inflammation. They also have anti-inflammatory properties, support cognitive function, and improve cholesterol levels. Walnuts also contain heart-healthy omega-3 fatty acids. Another hack I've used, especially when I'm in the midst of a fast, is consuming no-calorie sparkling water since the carbonation helps fill my stomach and helps me to feel satiated.

Consuming Enough Protein

Protein is an essential building block of life. You need it for building and repairing tissues like muscles, bones, and skin, as well as for producing hormones and enzymes, transporting nutrients and oxygen, supporting immune function, and providing energy to your cells.

In my podcast with Dr. Hyman, he noted, "Some believe we should limit protein (especially animal protein) and amino acids to silence mTOR and activate autophagy." Before moving on, let's break down a few of these terms. mTOR, which stands for "mechanistic target of rapamycin," is like your body's growth switch. (We'll encounter rapamycin later in the "Medications and Supplements" chapter.) When mTOR is active, it tells your cells to grow and divide. Autophagy, on the other hand, is your body's cellular clean-up crew ("auto" means self, and "phagy" comes from the Greek word for eating). It's the process where your cells break down and recycle old, damaged parts.

Some researchers think that eating less protein, especially from animal sources, can turn off mTOR and kick autophagy into high gear, potentially slowing down aging. However, as Dr. Hyman points out, the data isn't so clear-cut. When we completely abstain from animal proteins, we risk keeping mTOR "silenced for long periods," which means "we can't create new proteins or build muscle." It's like trying to renovate your house without bringing in any new materials. For maximum longevity, we need to find a balance. We want to activate the anabolic (building up) mTOR pathway enough to maintain and repair our tissues, but we also need to stimulate the catabolic (breaking down) AMPK pathway, which promotes autophagy and cleans up all the cellular junk. Think of it as alternating between periods of construction and cleaning up your body. The key is

Great options for breakfast, high in protein, low in sugar.

finding the right rhythm between these two processes—building up and breaking down—to keep our bodies in top shape as we age.

In 2023, I was able to add ten pounds of additional muscle mass to my frame, and from 2024 onward, my goal has been to maintain that additional muscle. To accomplish my goal, I significantly increased my protein intake. Typical advice suggests we should consume 0.80 grams of protein per pound of body weight (1.6 grams per kg). But what I've read suggests that's just too low, especially as we need to maintain muscle mass as we age, so now, my goal is to consume 1.0 grams of protein per pound of body weight. I take in a whopping 150 grams per day. (I weigh about 145 pounds.)

Typically, I eat protein from many of the following sources:

- Fish (wild salmon) three times per week, eggs, turkey, or chicken. (Again, I avoid tuna, swordfish, and large fish like the plague because they are high in mercury.)
- Lentils, chickpeas, and black, kidney, and pinto beans are great protein sources.
- Protein shakes three to five days per week. On the days I'm lifting weights, I'll consume a high-quality protein product. I typically alternate between a plant protein shake (I love the chocolate-flavored Ka'Chava shake) and a high-quality, grass-fed whey protein shake.
- Nutri11 is another protein drink (served hot) that I consume on my weight workout days, often as a coffee replacement. It has zero sugar (sweetened by monk fruit) and 11 grams of protein, and I love the taste.
- Green peas are another great protein source. One cup (160g) of cooked green peas contains 9 grams of protein.
- Almonds and sunflower, flax, and chia seeds are good protein sources.

Note: Don't try to consume a day's worth of protein all in one sitting; instead, spread it out over three to four servings throughout the day. Doing this will maximize muscle protein synthesis, enhance recovery, sustain an anabolic state, optimize nutrient utilization, and manage appetite.

During the four days of the week when I'm doing a heavy-weight workout and trying to consume maximum protein, here's an example of what I'll consume on a typical workout day:

Item	Grams of Protein	Calories
Whey Protein Shake	25 grams	120
Almond Milk	04 grams	160
Nutri11 Drink	11 grams	145
Blueberries (1 cup)	--	84
Walnuts (1/2 cup)	8 grams	300
Salmon (6 ounces)	34 grams	354
Broccoli (2 cups)	05 grams	60
Ka'Chava shake	25 grams	120
Almond Milk	02 grams	80
Chicken breast	43 grams	230
TOTAL	157 grams	1,653

What I Do NOT Drink

Equally important to what I do and do not eat is what I do and do not drink. Here is how I think about each of these drink options.

Sodas

I've eliminated 100 percent of sodas from my diet, given the added sugar and phosphoric acid they contain. Both regular and diet sodas are detrimental to overall health. A single 12-ounce can of soda typically contains about 39 grams of sugar, which is equivalent to approximately 9.3 teaspoons of sugar. This amount of sugar far exceeds the daily recommended limit set by health organizations. Diet sodas often contain artificial sweeteners like aspartame, which have been associated with an increased risk of metabolic syndrome, disrupted gut microbiota, and even potential links to cardiovascular disease and stroke. Beyond sugar, the phosphoric acid in both regular and diet sodas can erode tooth enamel and lead to dental problems and an imbalance in the body's calcium-to-phosphorus ratio, which is crucial for maintaining strong bones. Moreover, excessive consumption of phosphoric acid can increase the risk of kidney disease.

Fruit Juices

I've eliminated 100 percent of high-fructose fruit juices, which can spike my blood sugar. Fructose, especially in liquid form, can increase uric acid levels, blood pressure, and appetite. The bottom line is if you like oranges, eat a whole orange. The key benefit of eating a whole orange lies in its fiber content. Fiber slows down the absorption of sugar into the bloodstream, helping regulate blood sugar levels and promoting digestive health.

Alcohol

My father's favorite saying was "Pan Metron Ariston" (Παν μέτρον άριστον), which translates to "everything in moderation" and is attributed to the ancient Greek philosopher Cleobulus of Lindos, one of the Seven Sages of Greece. But when it comes to alcohol, the question is, how much is too much? I've eliminated almost all of my alcohol intake, save for an occasional glass of red wine. (I love Amarone.) On a biochemical basis, alcohol truly has very little medicinal benefit and is a major driver of microbiome disruption and leaky gut. The challenge, of course, is that drinking alcohol, which is typically an evening and late-night event, can also disrupt sleep and, over time, cause insomnia by interfering with the body's system for regulating sleep. Even though alcohol can initially cause us to be sleepy, it will disrupt ongoing sleep.

That being said, alcohol has positive side effects that are worth acknowledging. Many people who live long, healthy lives often also consume alcohol (in moderation). While the alcohol itself may be detrimental to some aspects of our health, the social interactions that typically take place while sharing a glass of wine (or your drink of choice) can be beneficial to mental and emotional health. More quality time spent with friends leads to a happier and more fulfilled life.

What I DO Drink

Water

I aim to drink 2+ liters of water daily, prioritizing fresh spring or sparkling water. In the morning, I'll drink two cups of water first thing on an empty stomach. The benefits are many, including replenishing fluids lost overnight, kickstarting hydration for the day, stimulating your metabolism, helping flush out toxins, enhancing cognitive function, preparing your digestive system for the day, and promoting regular bowel movements. On days when I'm focused on aerobic exercise, I'll mix LMNT electrolyte powder into a liter of water to replenish essential electrolytes, including sodium, potassium, and magnesium. I love their chocolate flavor.

Coffee and Other Morning Hot Drinks

In medical school, I would average five or more cups of coffee per day, always black without cream or sugar. I recently learned from my genetics that I'm a slow metabolizer of caffeine (much to my surprise!), and getting more than a cup of caffeinated coffee per day probably isn't in my best interest. So, these days, I typically switch to decaffeinated after my first cup of coffee.

While most of us think of coffee as a stimulant, it actually has several health-related benefits independent of caffeine. Coffee is one of the largest sources of antioxidants in the typical diet. These antioxidants, such as chlorogenic acid and polyphenols, help combat oxidative stress in the body, reducing the risk of developing and dying from chronic diseases like heart disease, cancer, and diabetes. Regular coffee consumption has been associated with a lower risk of neurodegenerative diseases such as Alzheimer's and Parkinson's. However, if you are a slow caffeine metabolizer like me, the caffeine can increase your heart rate and cause jitteriness. There is even some evidence that drinking a lot of caffeine in slow metabolizers can increase your risk for heart disease.[10]

As an alternative to coffee, I will typically have a mug of hot Moroccan mint tea (a green tea) that I will sweeten with 100 percent pure monk fruit extract (no calories and a taste I enjoy). In addition, I will occasionally enjoy MUD\WTR, a black tea powder containing a number of beneficial mushrooms, or a hot mug of Nutri11 (high in protein with no sugar or carbs).

The Surprising Benefits of 3–5 Cups per Day

A September 2024 article published in *Medscape* shared surprising outsized benefits of significant coffee consumption. Titled "Coffee's 'Sweet Spot': Daily Consumption and Cardiometabolic Risk," the article summarized a comprehensive study published in the *Journal of Clinical Endocrinology and Metabolism*, leveraging data from the UK Biobank.[11] The study found that consuming three to five cups of coffee per day could lead to a 15 percent reduction in cardiovascular risk compared to non-drinkers. Intriguingly, the research went beyond mere association, delving into the metabolomic effects of coffee consumption. Analysis of 168 individual metabolites showed

that coffee intake significantly altered 80 of these, potentially explaining its cardioprotective effects. Notably, coffee consumption was linked to lower levels of VLDL cholesterol and saturated fatty acids, both known as risk factors for heart disease. The study also found similar benefits for tea and caffeine intake, suggesting that the positive effects may not be limited to coffee alone. While the research presents compelling evidence for coffee's health benefits, it's important to note the distinction between black coffee and high-calorie, sugary coffee drinks. This study adds to a growing body of evidence supporting moderate coffee consumption as part of a healthy lifestyle.

How & When I Eat: Fasting, Timing, and Calories

Slowing Down, Breathing Deep, and Vitamin "O"

Have you ever heard of vitamin O? Dr. Helen Messier calls this taking a deep breath of oxygen at the start of a meal to activate your parasympathetic ("rest and digest") system and eating slowly and mindfully. You can feel your entire nervous system relax when you take three deep breaths through your nose. Give it a try. In this state, you're putting your digestive system in the ideal state for receiving nutrients. You're telling your body it's time to relax and fuel up. Unfortunately, in this fast-paced, 24-7 world, most of us are doing the opposite: eating on the run, chowing down food while driving, answering emails, or, worst of all, eating while watching the dystopian evening news on TV.

Eating under stressful situations activates your sympathetic nervous system, which is the worst way to digest your food. During a sympathetic state, the body is in "fight or flight" mode, characterized by increased heart rate, elevated stress hormones, and reduced blood flow to the digestive organs. This state impairs digestion and nutrient absorption, diverting energy away from the gastrointestinal system and toward dealing with perceived threats.

To maximize both enjoyment and the healthy absorption of nutrients and fully digest your meal, activating your parasympathetic system is critical. To do this, take a few deep breaths to slow down your heart rate and increase your oxygen intake at the start of any meal. This is why, in many cultures, people say grace or share what they are grateful for at the beginning of a meal. Activating and increasing your parasympathetic system stimulates the production and release of digestive enzymes and gastric juices, which leads to better digestion and nutrient absorption. It also causes blood flow to be directed toward the digestive organs.

Meal Sequencing: The Order in Which You Eat Your Food Matters (A LOT)

In her *New York Times* bestselling book, *Glucose Revolution: The Life-Changing Power of Balancing Your Blood Sugar,* Jessie Inchauspé discusses a simple but powerful concept called meal sequencing. Picture this: You're seated at your dinner table, a colorful plate of food before you, and you're about to take the first bite. What if I told you that the order in which you eat your meal could have a significant impact on your health, specifically in managing blood sugar levels and supporting weight control?

Imagine your plate divided into three parts: carbohydrates, protein, and vegetables (fiber). Now, let's arrange them in a way that can transform your eating experience and impact health and weight gain.

First in Line – Fiber-Packed Vegetables: Begin your culinary journey with the vibrant, fiber-filled bounty of non-starchy vegetables. Think kale, broccoli, bell peppers, and spinach. The more colorful, the better. Starting with your vegetables serves two important objectives. First, it kick-starts your meal with a burst of nutrients and hydration, making you feel more satisfied from the get-go. Second, the fiber in vegetables helps slow the absorption of the other foods. This means you won't experience those rapid spikes in blood sugar levels that can lead to energy crashes and unwanted cravings. Even though vegetables are technically carbohydrates, their fiber content makes them what we call resistant starches. This means they resist digestion in the small intestine, where we absorb our food, and are instead fermented in the large intestine, where they can feed our essential microbiome.

Next Up – Protein Power: After your veggies, enjoy a succulent piece of grilled chicken, fish, or a hearty serving of plant-based tofu or legumes. At this stage, protein takes center stage in stabilizing your blood sugar levels and curbing your appetite. Protein-rich foods provide a sense of fullness and satisfaction that helps you control portion sizes. Plus, they work in harmony with the fiber from your vegetables to create a dynamic duo for regulating blood sugar.

Last Up – Carbohydrates: Finally, if you eat any starchy carbohydrates, like pasta and bread, please do that last. Eating your carbohydrates last allows your body to prioritize the digestion of fiber and protein, reducing the potential for blood sugar spikes.

Time-Restricted Eating and Caloric Restriction

One of the most hotly debated areas around diet has to do with terms like "intermittent fasting," "time-restricted eating," and "caloric restriction"—basically, how much you should eat and when.

When you ask longevity experts like David Sinclair, PhD, and Eric Verdin, MD (CEO of the Buck Institute), they'll tell you that eating less is one of the most powerful ways to extend healthspan and lifespan.

Research consistently shows that mice and rats subjected to a caloric restriction (CR) diet live significantly longer than their counterparts who were allowed to eat freely. These animals not only live longer but also show delayed onset of chronic diseases such as cancer, cardiovascular disease, and neurodegenerative disorders. For instance, a landmark study published in *Nature* found that mice on a CR diet lived up to 40 percent longer than those on a standard diet, with fewer incidences of age-related diseases.[12]

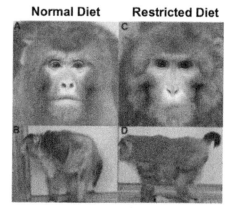

(Image from "Caloric restriction delays disease onset and mortality in rhesus monkeys" by Ricki J. Colman, et al., 2009.)

Closer to humans is a well-known study conducted by the National Institute on Aging on rhesus monkeys.[13] In this study, monkeys fed just 30 percent less monkey chow (i.e., 30 percent fewer calories) showed a reduced incidence of diabetes, cardiovascular disease, and cancer.

At the cellular level, caloric restriction has been shown to activate various biological pathways associated with longevity. One key pathway is the reduction of insulin-like growth factor 1 (IGF-1) signaling, which is known to influence aging and disease processes. Lower levels of IGF-1, as seen in CR, have been linked to increased cellular repair mechanisms, enhanced autophagy, and reduced oxidative stress—all of which contribute to healthier aging. Additionally, caloric restriction has been found to upregulate sirtuins, a family of proteins that play a critical role in regulating cellular health, DNA repair, and metabolic efficiency, further supporting its role in promoting longevity.

Human studies, while less extensive, also provide promising insights. The CALERIE (Comprehensive Assessment of Long-term Effects of Reducing Intake of Energy) trial, a well-known clinical study, examined the effects of a 25 percent reduction in caloric intake in non-obese humans over two years.[14] The results showed improvements in markers of cardiovascular health and insulin sensitivity and reduced oxidative stress, all of which are associated with increased lifespan and reduced risk of chronic diseases.

In summary, the scientific evidence strongly supports the idea that eating less, through caloric restriction, can be an effective pro-longevity protocol. By reducing caloric intake, individuals may extend their lifespan and improve their overall health, delaying the onset of age-related diseases and enhancing the quality of life as they age.

Professor Sinclair puts it this way: "The most important eating habit for longevity is to eat less often. It's not just what you eat. It's also when you eat, and this constant eating of three meals a day plus snacks is making us age faster than we need to. I like to eat within a period of about six hours a day. Over time, I learned to skip meals. I'm not always successful. Sometimes, I have breakfast when I'm in a beautiful place, but my goal is to not eat a large meal until dinner, and then I eat a very healthy vegan meal.

"So what's the science behind the benefit of the time-restricted feeding? If you're down to one meal a day, which I am now, you shed weight, and then you get your 20-year-old body back—that's a nice bonus. It's the period of not eating that's so important for boosting the body's defenses against aging to maximize longevity. But these long-extended periods are doing a real deep cleanse on the body and turning on that autophagy, the process of recycling proteins very deeply. There is a set of genes called sirtuins that get turned on when there's not enough energy in the body. So, if you don't have a lot of sugar in your bloodstream or a lot of protein, they'll get turned on, and they defend the body against the damage that causes aging."

However, the data is, unfortunately, not that simple. In Peter Attia's book *Outlive*, he reports on a 2020 clinical trial known as the TREAT (Time-Restricted Eating for Weight Loss) study (published in *JAMA Internal Medicine*) in which 116 overweight or obese adults were divided into two groups: one group followed a time-restricted eating (TRE) regime where they ate within an 8-hour window (from 12:00 p.m. to 8:00 p.m.), and the other group followed a consistent meal timing (CMT) pattern, eating three structured meals per day.

Peter with David Sinclair and Nir Barzilai. David Sinclair, PhD is a Professor at Harvard Medical School. (He doesn't look 55 years old, does he?)

The study's primary goal was to determine if the 16:8 TRE regimen led to greater weight loss and improvements in cardiometabolic health compared to the CMT group. Over the twelve weeks, both groups experienced a modest weight reduction, but the difference between the two groups was not statistically significant.

There was, however, serious and, in my opinion, valid criticism of the trial, arguing that the study only tested a single TRE window (12:00 – 8:00 p.m.) and did not explore other fasting windows that might align better with the body's circadian rhythms. Research suggests earlier eating windows (such as 8:00 a.m. – 4:00 p.m.) might be more effective for weight loss and metabolic health. Second, the study did not directly measure participants' caloric intake, leaving uncertainty about whether the TRE group actually consumed fewer calories than the control group. Third, the 12-week duration of the study is considered too short by some experts to fully assess the long-term effects of TRE on weight loss and metabolic health.

So, what should we make of this data? How should we think about this? I go back to our evolutionary past. For millions of years, our ancestors lived as hunter-gatherers, often experiencing many periods of food scarcity, often interspersed with periods of abundance. This lifestyle meant humans evolved to thrive in environments where food was not always readily available, and fasting was a natural part of life. This evolutionary background suggests that our bodies are well-adapted to periods of caloric restriction and intermittent fasting, potentially explaining why these practices can lead to health benefits, such as improved metabolic function, increased insulin sensitivity, and enhanced cellular repair mechanisms through processes like autophagy.

Our challenge is that the constant availability of high-calorie, processed foods combined with our primarily sedentary lifestyle diverges sharply from the environments in which our ancestors evolved.

My Eating Strategy

I've adopted a number of strategies that work for me. First and foremost, my goal is to eat mindfully and, in so doing, to reduce my overall caloric intake. This plan includes four key things: first, drinking a full glass of water at the beginning of my meal; second, limiting how much I put on my plate and resisting the desire to get seconds until my stomach has caught up with my brain; third, eating slowly and enjoying every bite; and fourth, chewing every bite ten or more times.

In addition, I have chosen to participate in time-restricted eating, but I do this in coordination with my workout routines (see Chapter 2 on "Exercise" and Chapter 7 on "Routines").

TRE Days (18 Off, 6 On): Monday, Tuesday, and Friday

Most people participating in time-restricted eating tend to skip breakfast and eat a normal lunch and dinner, thinking that the total length of the fast is more important than when the fasting period occurs. Most of us have more willpower in the morning to skip a meal and tend to enjoy a normal dinner time with friends and family—which is definitely the case for me.

In reality, eating breakfast and lunch and skipping dinner is better for controlling our metabolism and more aligned with our clock genes since we are most insulin-sensitive in the morning and more insulin-resistant in the evening.

The bottom line is a 6-hour "eating period" is healthier (from a metabolic perspective) between 8:00 a.m. and 2:00 p.m. rather than 2:00 p.m. and 8:00 p.m. If you can do this, more power to you. I have two thirteen-year-old boys and desire to dine with them, so skipping dinner has proven challenging. Having said that, I've shifted my family dinner time earlier. We now shoot for 5:00 p.m., giving me a solid four-hour fasting period before sleep and a feeding period typically starting at noon and ending by 6:00 p.m.

As such, on the days that I'm not working out (i.e., not consuming significant protein before/after weight work), I'm endeavoring to eat between noon and 6:00 p.m., and yes, occasionally, I'll go until 7:00 p.m.

- **Avoiding Food After 6:00 p.m.:** Because I set my bedtime at 9:30 p.m., it's critical for me to avoid any food or snacks after 6:00 p.m. (7:00 p.m. at the latest). This offers a two or three-hour buffer period before sleep. Recent studies have highlighted the profound impact of consuming even a small amount of sugar before you go to bed.[15] Eating just one gram of sugar an hour before sleep can impact gene expression and ruin the benefits of time-restricted eating. Research from the Salk Institute has shown that introducing any caloric intake during periods meant for fasting can effectively "turn off" the benefits of TRE by desynchronizing the body's circadian clocks. This desynchronization can lead to impaired glucose metabolism, reduced fat burning, and increased risk of metabolic disorders, thereby negating the health benefits associated with fasting periods.

- **Morning Fasting (Waking – Noon):** On days that I'm focused more on aerobic, Zone-2, and interval (VO2 Max) training, I consume water, coffee, and green tea. Keeping busy and enjoying sparkling water allows me to get through that fasting period relatively easily.

I have more physical and cognitive energy these mornings because blood is not being diverted to my digestive system to digest a large breakfast.

Workout Days (12 Off, 12 On): Wednesday, Thursday, Saturday, and Sunday

When building muscle and engaging in a weight workout, I aim to consume sufficient protein before and after the workout with an overall protein goal for the day of 1 gram per pound of body weight, which, for me, is 150 grams.

On these days, I still try to support a 12-hour "on" and 12-hour "off" fasting routine. As such, I still avoid all foods after 7:00 p.m. and don't have anything again until after 7:00 a.m. My first hour after waking is limited to water, coffee, or green tea. After that, I'll consume a protein shake, as described earlier in this chapter. I'll expand on my routine in the next chapter on exercise.

What About Longer Fasts?

In recent years, multi-day fasting and adopting fasting-mimicking diets (FMDs), such as the ProLon diet, have gained substantial attention in longevity research.

Cellular Repair and Autophagy: One of the most compelling benefits of multi-day fasting is its ability to enhance autophagy—a natural process by which cells remove damaged components and regenerate new ones. Autophagy is critical for maintaining cellular health and preventing the accumulation of dysfunctional proteins and organelles associated with aging and diseases like Alzheimer's. A study published in *Nature* highlighted that fasting triggers autophagy across various tissues, including the liver, muscle, and brain, thereby supporting improved cellular function and longevity.[16]

Improved Metabolic Health: Fasting has also been shown to significantly affect metabolic health. Research published in *Cell Metabolism* demonstrated that periodic fasting can improve insulin sensitivity and stabilize blood sugar levels, reducing the risk of type 2 diabetes and other metabolic disorders.[17] Even without weight loss, fasting was found to reduce visceral fat and improve metabolic markers, underscoring its benefits beyond mere calorie reduction.

The ProLon Fasting Mimicking Diet: Developed by Dr. Valter Longo and his team at the University of Southern California, the Prolon FMD is designed to replicate the benefits of fasting while allowing for minimal calorie intake. Clinical trials have shown that ProLon can effectively reduce IGF-1, a growth factor associated with aging and cancer while promoting cellular regeneration. Participants in these studies experienced weight loss, improved metabolic markers, and reduced risk factors for age-related diseases. Prolon is a 5-day program that is relatively painless and something I've been doing twice per year. My longevity physician tells me that on an annual basis, I would ideally do this once a month for three consecutive months.

Peter loves Prolon as a 5-day fast that he does multiple times per year. It's affordable, easy, and it works.

Using a Continuous Glucose Monitor (CGM): Measuring My Blood Glucose

Peter Drucker's famous saying, "If you can't measure it, you can't manage it," applies equally well here. To understand what foods do and do not spike my blood sugar, I use a continuous glucose monitor or CGM. There are two leading CGM devices: Dexcom G7 and FreeStyle Libre 3 (by Abbott Labs). Both provide a continuous wireless readout to your smartphone where you can examine a minute-by-minute chart and determine exactly what eating those M&Ms is doing to you. I typically use a Dexcom G7 with an app called Levels that provides real-time data from my Dexcom G7, allowing me to immediately see how my blood glucose levels respond to different foods and activities. The app also analyzes your glucose data and offers personalized recommendations on diet and lifestyle changes to optimize your metabolic health. Using a CGM helps keep me aware of my eating habits in the same way that my Oura Ring helps me monitor my sleeping habits.

The GLP-1 Revolution: Miracle Cure or Metabolic Gamble?

No health or longevity book published today would be complete without discussing GLP-1 receptor agonists, which are taking the world by storm. These drugs, with brand names like Ozempic and Wegovy, have been hailed as game-changers in the fight against obesity and type 2 diabetes. However, as with any powerful tool, they come with promise and peril. Let's take a look.

The GLP-1 Basics: Your Body's Natural Appetite Suppressant

Before exploring the drugs, let's understand how GLP-1 (or glucagon-like peptide-1) receptor agonists work. GLP-1 is produced by L cells in the digestive tract in response to food intake. It acts as a natural appetite suppressant, signaling your brain that you're full, saying, "Hey, we've eaten enough. Time to put down the fork!" GLP-1 helps regulate blood glucose levels and promotes insulin secretion. In essence, GLP-1 is your body's built-in portion control and metabolic regulator.

The Pharmaceutical Leap: Supercharging GLP-1

Now, here's where things get interesting. Pharmaceutical researchers had a thought: "What if we could amplify this natural process?" Their solution? Create drugs that mimic GLP-1 but at levels far beyond what your body typically produces; we're talking up to 100 times the normal amount, and the results were nothing short of dramatic:

- Appetites vanish
- Pounds melt away
- Blood sugar levels stabilize

It seemed like the holy grail of weight loss had been discovered. But as we'll see, this powerful intervention comes with its challenges.

Peter uses a Dexcom G7 & Levels app to monitor avg blood glucose and avoid foods that cause spikes. At 6:45 a.m., Peter worked out.

The Promise: Unprecedented Weight Loss

The most obvious benefit of GLP-1 drugs is their effectiveness in promoting weight loss. Many users report shedding pounds that had stubbornly resisted diet and exercise for years. For those struggling with obesity-related health issues, this can be life-changing.

Beyond weight loss, these drugs offer other potential benefits:

- Improved blood sugar control for type 2 diabetics
- Reduced risk of cardiovascular events in high-risk individuals
- A reprieve from the constant battle with food cravings

The Perils: Unintended Consequences

However, as the saying goes, "There's no such thing as a free lunch." GLP-1 drugs come with three principal drawbacks.[18] The first challenge is muscle loss. One alarming finding is that a significant portion of the weight loss may come from muscle, not just fat. As discussed in this book, muscle mass is critical to longevity. The second downside is the rebound effect. What happens when you stop taking the drug? Research shows that about 70 percent of the lost weight comes back—and it's mostly fat. This means you could end up with less muscle and more fat than when you started, a metabolic double-whammy. And the third drawback is the side effects, which include nausea, vomiting, and digestive issues—common companions on this weight loss journey. While usually manageable, they can be quite unpleasant.

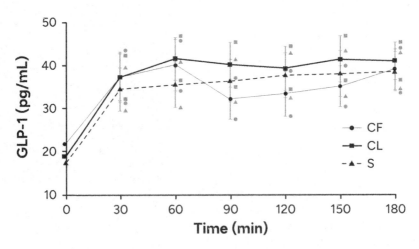

Key for Graph Above:
CF (Solid line) means Carbs First (GLP-1 drops after 60 minutes), CL (Dotted line) means Carbs Last (GLP-1 stays up for at least 3 hours), S (Dashed line) means all foods eaten together

Navigating the GLP-1 Landscape

If you're considering GLP-1 drugs or are already taking them, here are three strategies to maximize benefits and minimize risks. First is weightlifting. Resistance training is your new best friend. Aim for at least three sessions a week to preserve that precious muscle mass. The second is sufficient protein consumption. Target 1 gram of protein per pound of body weight daily. And third, build healthy habits. Use your time on this drug to build sustainable eating and exercise routines. The drug should be a tool, not a crutch, in your health journey.

How to Build Your GLP-1 Naturally

What if I told you that you could increase your GLP-1 levels without a prescription? Here are some evidence-based strategies:

1. **The Sequence Secret**: As mentioned earlier in this chapter, eat your food in the following order: vegetables first, then proteins and fats, and starches and sugars last. This simple switch can boost your post-meal GLP-1 levels by up to 38 percent.

    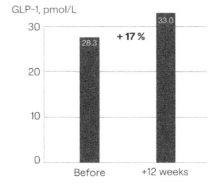

    200mg Eriomin tested against a placebo. (Clinical trial PMID 3576695)

2. **Chew Your Food (A Lot)**: Take your time and chew thoroughly—at least ten times per bite. Your L cells need time to sense the food and release GLP-1. How often do you get overstuffed because you don't feel full until later?

3. **Yerba Maté Magic**: This South American beverage isn't just a tasty caffeine alternative; it's also a GLP-1 booster.

4. **Lemon Power** (Eriocitrin): The lemon extract in Eriomin tells your L-cells to produce more GLP-1. Concentrated supplements like Eriomin have shown promising results in clinical trials.

5. **Move Your Body**: Regular exercise, especially high-intensity interval training, can naturally increase GLP-1 levels.

Insights from Dean Ornish, MD: The Power of Lifestyle Medicine

We mentioned earlier that your genes only account for 7 to 30 percent of your longevity. That means your lifestyle has a profound effect on your health. But can your lifestyle alone reverse heart disease, extend your telomeres, and reverse dementia? Dean Ornish, MD, the renowned physician, researcher, and tireless pioneer in lifestyle medicine, has demonstrated that the answer is a resounding *yes*.

Dr. Ornish has been a dear friend for nearly two decades, and he is someone whose work I wish to celebrate in this book. I've added this discussion on his research at the end of the chapter on "Longevity Diet" to introduce these concepts early. However, Dean's work could have easily also gone into the chapters on exercise and mindset.

Ornish, who many consider the father of lifestyle medicine, has repeatedly demonstrated that comprehensive lifestyle changes not only prevent but also reverse chronic diseases in areas such as coronary heart disease and even dementia. He has written multiple best-selling books on the topic, including his latest *Undo It!: How Simple Lifestyle Changes Can Reverse Most Chronic Diseases*, co-authored by Anne Ornish. Dean's 45 years of peer-reviewed research has been published in the most prestigious medical journals, and in 2024, that culminated in a CNN documentary produced by Dr. Sanjay Gupta and broadcast on the *Anderson Cooper Show*.

Ornish's approach, often referred to as the Ornish Lifestyle Medicine Program, focuses on four key areas:

1. A low-fat, whole-food, plant-based diet low in sugar and refined carbs
2. Regular exercise
3. Stress management techniques such as meditation
4. Social and community support

In his landmark study published in 1990 in *The Lancet*, Ornish showed for the first time that patients with severe coronary artery disease who followed his program experienced a reduction in chest pain by 91 percent within just weeks.[19] More impressively, after one year, 82 percent of patients in the lifestyle change group experienced *reversal* of their coronary atherosclerosis, whereas the control group recorded a 53 percent *progression* in the disease.

In a follow-up study published in the *Journal of the American Medical Association* in 1998, Ornish demonstrated that these lifestyle changes could produce sustained improvement over a five-year period.[20] Patients following his program showed a 7.9 percent relative improvement in coronary artery blockages after five years, while those in the control group experienced a 27.7 percent relative worsening.

Ornish's work extends beyond heart disease. In studies published in 2008 and 2013, he showed that these lifestyle changes could actually increase telomerase activity by 29 percent in just three months and, for the first time, significantly lengthen telomeres after five years.[21] Telomerase is an enzyme that repairs and lengthens telomeres (the specialized DNA sequences located at the ends of chromosomes), which are associated with longevity. Furthermore, Ornish's program has been shown to impact gene expression. In a 2008 study in the *Proceedings of the National Academy of Science*, he found that after just three months on his program, over 500 genes were favorably changed in their expression, including genes that help prevent cancer and other chronic diseases.[22]

The success of Ornish's approach has led to its adoption by many healthcare providers and insurance companies. In 2010, Medicare began covering "Dr. Dean Ornish's Program for Reversing Heart Disease" as an intensive cardiac rehabilitation program, making it more accessible to millions of Americans, now offered virtually via Zoom.

Building on this, Ornish's latest research, published in 2024 in *Alzheimer's Research & Therapy*, marks a significant breakthrough in Alzheimer's treatment.[23] In a randomized control trial, patients with mild cognitive impairment or early dementia due to Alzheimer's who followed Ornish's intensive lifestyle intervention showed, for the first time, statistically significant improvements in cognition and function after just twenty weeks. The intervention group demonstrated significant differences in three out of four standard cognitive tests used in FDA drug trials, with the fourth test showing borderline significance.

Key findings from the Alzheimer's study include:

1. Seventy-one percent of patients in the intervention group improved or remained stable, while 68 percent in the control group worsened, and none improved.
2. Patients reported regaining lost cognitive abilities, such as reading comprehension and financial management skills.
3. A dose-response correlation was observed between the degree of lifestyle changes and improvements in cognition and function.
4. Significant improvements were seen in key blood-based biomarkers, including the $A\beta42/40$ ratio, a measure of amyloid in Alzheimer's disease.
5. The intervention group showed beneficial changes in gut microbiome composition related to Alzheimer's risk.

Today, Ornish's research continues to demonstrate that lifestyle medicine can be a powerful tool in preventing and often reversing many chronic diseases, potentially offering a more accessible and cost-effective approach than traditional pharmaceutical interventions. I urge you to check out Dr. Ornish's work and programs at Ornish.com.

Chapter 2
Exercise Blueprint: Build Strength, Endurance, & Longevity

"The only bad workout is the one that didn't happen."
—**Unknown**

*"Exercise is king. Nutrition is queen.
Put them together, and you've got a kingdom."*
—Jack LaLanne

In 2006, Jacinto Bonilla celebrated his 66th birthday in an unusual way—by joining his first CrossFit class.[24] A former Marine and lifelong fitness enthusiast, Jacinto was no stranger to physical challenges. But even he couldn't have predicted how this decision would transform his golden years. Fast forward to 2019, and Jacinto, now 78, was still going strong, regularly outperforming people half his age in grueling workouts that combined weightlifting, cardio, and functional movements.

Jacinto Bonilla goes hard on exercise.

Jacinto's story caught the attention of researchers and fitness enthusiasts alike. Here was a man in his late seventies, deadlifting over 300 pounds, performing muscle-ups, and competing in high-intensity fitness competitions. However, what truly set Jacinto apart wasn't just his physical prowess; it was his vitality, energy, and zest for life. While many of his peers were grappling with age-related decline, Jacinto was thriving, proving that chronological age doesn't have to dictate biological age.

The science backs up Jacinto's experience. Studies have shown that resistance training, like the weightlifting Jacinto incorporates into his routine, can significantly slow down and even reverse aspects of aging.[25] It preserves muscle mass, maintains bone density, improves metabolic health, and enhances cognitive function. Jacinto's journey isn't just inspiring; it's a powerful testament to the role of exercise, particularly strength training, in extending the quantity and quality of our years. As Jacinto often says, "Age is just a number. What matters is how you feel and what you can do."

"Exercise is the single most important pro-longevity activity you can undertake."

If you are over 60 years old, exercising just twice per week with weights (resistance) and doing push-pull exercises (push-ups, pull-ups, deep knee bends, etc.) has the effect of reducing all-cause mortality by 50 percent and reducing your risk of cancer three-fold.[26]

to become part of your regular routine since muscle decreases approximately 3 to 8 percent per decade after the age of thirty, and this declines at an even higher rate after the age of sixty.

What I Do to Build Muscle

Heavy Weight Workout

I work out at least four times per week using heavy weights for both my upper and lower body. Each session is roughly 40 to 50 minutes long and, when possible, guided by a trainer who pushes me harder than I can typically push myself. The only real disruption to this routine is travel. One way that I try to overcome the challenge of working out during travel is to invite others traveling with me to meet me in the gym. Making an appointment with a colleague for a 7:00 a.m. workout at the gym is one way to dismiss the multitude of rationalizations and excuses you can come up with lying in bed the next morning.

Tonal Workout

To gain and maintain muscle mass this year, I invested in an at-home all-in-one AI-enabled workout system called Tonal, which I use as an alternative to free weights. The Tonal System strives to marry the best of strength training with the innovation of smart technology. At its core, Tonal is a wall-mounted unit that harnesses electromagnetic resistance to simulate weights, allowing users to engage in a wide variety of strength training exercises without traditional bulky gym equipment. Accompanied by an interactive display, the system intelligently adapts to each individual's needs, calibrating workouts in real-time and providing expert coaching through guided routines. The importance of the Tonal System lies not just in its space-saving design but in its ability to deliver personalized, efficient, and dynamic workouts. It redefines the boundaries of traditional training by offering a tailored experience that can accelerate muscle growth, enhance strength, and revolutionize the way we understand home fitness. One final benefit of Tonal is it is likely safer than lifting free weights, reducing the risk of injuries caused by incorrect form.

Creatine

Building muscle goes beyond just lifting weights; it also requires the proper diet and supplementation. Beyond consuming 150 grams of protein (for my body

So, what type of exercise should you strive to incorp[orate,] often, and how much? When I think of exercise, I'm [guided by] two objectives: My first goal is to maintain and, if po[ssible, grow my] mass, and my second is to optimize my metabolic and [cardio health.] Below, we'll dive into both of these; I'll share my exercis[e routine and] why exercise is so important as a pro-longevity practic[e.]

Increasing Muscle Mass

Adding muscle to your frame is fundamental for healt[h. As I] mentioned earlier, I achieved my goal last year of add[ing ten] pounds of skeletal muscle (not an easy feat). Now, my go[al is to maintain it.]

Why? A number of studies have demonstrated a direct [correlation between] longevity and muscle mass. One study published in *The A[merican Journal of] Medicine* looked at data from over 10,000 people and fou[nd that those with] the highest muscle mass were 30 percent less likely to die [over the study] period than those with the lowest muscle mass.[27] Other [studies published] in *Gerontology* found that people with sarcopenia (a condi[tion characterized] by low muscle mass) had a *50 percent increased risk of death*[.]

One major benefit of increased muscle mass is its ro[le in preventing] injuries from falling. Many deaths occur when prolonged sarcopenia ultimately leads to a fall, resulting in a fractured hip or pelvis, followed by hospitalization, pneumonia, and ultimately death.

Note: According to a 2019 study in *Acta Orthopaedica*, in adults over the age of sixty-five, the one-year mortality rate after a hip fracture is 21 percent for those whose fracture is surgically repaired.[29] If the fracture is not repaired, the one-year mortality jumps to 70 percent!

For this reason alone, building and maintaining muscle mass is absolutely critical and is not a "once and done" activity. Building muscle

Peter does heavy weight workouts 3 or 4 times per week.

weight), I also supplement 5 grams of creatine. Research shows supplementing with creatine can double your strength and lean muscle gains when compared to training without it.[30] Creatine, a substance that naturally occurs in muscle cells, enhances muscle growth and strength in several ways: by increasing ATP production for high-intensity workouts, promoting cell signaling for muscle repair, boosting cell hydration for muscle size, stimulating protein synthesis, reducing protein breakdown, and lowering myostatin levels to increase muscle growth potential. Recently, I've tried a creatine monohydrate gummy called Create, which makes it easy to use when traveling. Otherwise, I mix a scoop of creatine powder into my protein shake. There are many brands. I'm currently using one made by Thorne.

Peter's recently installed Tonal System to make safe & impactful workouts happen!

Measuring Your Muscle Mass

So, how do you know that your weight workouts are adding muscle mass other than flexing in front of the mirror? One important way is to use technology to measure your gains. The gold standard is a Dexa Scan (which I get every year at Fountain Life), but a lower-cost at-home option is something like an InBody H20N smart weight analyzer. The H20N works by using what's called bioelectrical impedance. This smart scale sends a very low, safe electrical current through the body, measuring resistance (impedance) as it passes through different tissues of your body. It utilizes multiple frequencies, feeding the data to its algorithms to interpret the measurements, resulting in a detailed breakdown of body composition, including muscle mass, fat mass, water content, and more. However, it's worth noting that while devices like the InBody H20N are generally accurate, factors like hydration status, recent exercise, and food intake can affect results.

VO2 Max & Mitochondrial Health

Having plenty of muscle is great, but what about the metabolic health of your muscle and its mitochondria—in other words, the ability of your mitochondria (the power plants of your cells) to efficiently convert oxygen and glucose or fat into energy for your body? We measure that efficiency by using the VO2 max or maximal oxygen uptake. It measures the maximum amount of oxygen an individual can use during intense exercise.

It's a key indicator of our longevity. A higher VO2 max generally indicates a higher aerobic and endurance capacity.

VO2 max is important for several reasons, including (i) cardiovascular health, reducing the risk of heart disease and stroke; (ii) physical fitness and mobility; (iii) chronic disease prevention, reducing the risk of diabetes, hypertension, and certain types of cancer; (iv) mental health and cognitive function, reducing the risk of depression, anxiety, and cognitive decline, including conditions like Alzheimer's disease; and (v) longevity.

Several studies have found a direct correlation between VO2 max and longevity. Higher VO2 max levels in middle age have been associated with increased survival rates in older age.[31] "The relationship between VO2 max and all-cause mortality is quite good," says Dr. Michael Joyner, an anesthesiologist and fitness expert at the Mayo Clinic. "The odds of dying in the next ten years are markedly low if your VO2 max is high."

Because VO2 max changes with age and gender, below is a chart to help you evaluate where you stand in this important metric.

VO2 Max Values by Age and Gender

	Age	5th	10th	25th	50th	75th	90th	95th
MEN	20-29	29.0	32.1	40.1	48.0	55.2	61.8	66.3
	30-39	27.2	30.2	35.9	42.4	49.2	56.5	59.8
	40-49	24.2	26.8	31.9	37.8	45.0	52.1	55.6
	50-59	20.9	22.8	27.1	32.6	39.7	45.6	50.7
	60-69	17.4	19.8	23.7	28.2	34.5	40.3	43.0
	70-79	16.3	17.1	20.4	24.4	30.4	36.6	39.7
WOMEN	20-29	21.7	23.9	30.5	37.6	44.7	51.3	56.0
	30-39	19.0	20.9	25.3	30.2	36.1	41.4	45.8
	40-49	17.0	18.8	22.1	26.7	32.4	38.4	41.7
	50-59	16.0	17.3	19.9	23.4	27.6	32.0	35.9
	60-69	13.4	14.6	17.2	20.0	23.8	27.0	29.4
	70-79	13.1	13.6	15.6	18.3	20.8	23.1	24.1

(Percentile columns)

EXERCISE BLUEPRINT

How I Work on Improving My Mitochondria and VO2 Max

1. Steady-State Cardio: I aim to get at least three (ideally four) days each week of moderate "Zone 2" cardio exercise for 45–60 minutes. Zone 2 training has been shown to be the best way to improve the functioning and efficiency of your mitochondria. For me, this involves getting my heart rate to approximately 105–110 bpm and holding it steady through light jogging, brisk walking, cycling, or tennis. My go-to is a stationary bicycle, where I can take my board and team Zoom calls that otherwise would have me sitting around.

Here's how you calculate your Zone 2 heart rate.

First, calculate your "Max HR," which is equal to 220 minus your age.
For me, it's 220 − 63 = 157

Second, calculate the lower limit of Zone 2, which is equal to your Max HR x 60 percent.
For me: ~94 bpm

Third, calculate the upper limit of Zone 2, which is equal to Max HR x 70 percent.
For me: ~110 bpm

You can also do a lactate threshold test that will determine your exact Zone 2 heart rate, which I do at Fountain Life during my annual upload. This involves getting on a stationary bike, peddling at increasing speeds, and having a med tech sample your blood lactate levels from a finger prick at regular intervals. I plan to test this every year to track my performance improvements.

2. Interval Training Cardio: Beyond Zone 2, during my board

Peter using a Technogym stationary bike for Zone-2 cardio and HIIT workouts.

and staff meetings, I endeavor to focus on high-intensity interval training (HIIT) at least three days per week using my Technogym stationary bike. I warm up for five minutes, then alternate between one minute of high-intensity exercise (such as sprinting or fast cycling) and one minute of super-low-effort recovery. Currently, my goal is to get my bike to 175–200 watts of output and a heart rate of 140–145 bpm for one minute, followed by a very slow 25 watts of output where I'm barely peddling, during which my heart rate can recover. I repeat this cycle for fifteen minutes. This one kicks my butt. Luckily, the Technogym bike has an easy HR and power (Watts) readout, making this manageable.

Sitting Is the New Smoking

Most of us take our Zoom calls sitting on our butts, staring at our computer screens for hours at a time. We don't realize how much of our day we are physically idle. I think of "sitting as the new smoking"—something to be minimized.

During and after COVID, I've made it a habit of taking as many of my in-person meetings as possible as "walk and talk" sessions, which has helped me hit my daily goal of 10,000 steps or more as measured on my Apple Watch.

In addition, if my meeting is over Zoom, I can use my Lifespan walking desk or TechnoGym stationary bicycle to keep moving and work on Zone 2 training.

Stretching is essential to avoid injury.

Avoiding Injury

As we grow older, our biggest enemy is no longer sugar or even lack of sleep; it's *injury*. An injury can cause a sequence of events that can unravel all the pro-longevity work you've done. A fall that breaks any bone or tears a ligament can result in a prolonged period of bedrest that can wipe away many months of muscle accumulation or VO2 max improvements.

The antidote here is, first and foremost, building your muscle reserve, as discussed earlier. Beyond exercise, it's all about balance, flexibility, and stretching. As we get older, we tend to get stiffer and more unstable. Maintaining or gaining flexibility and working on proprioception, which is an awareness of the position and movement of your body, is critical in preventing injury. I must admit that flexibility is my Achilles heel, and I need to sign up for some yoga courses and stretching sessions.

Why Exercise Matters So Much

In a 2018 study published in *The Lancet*, researchers looked at the link between physical activity and mortality.[32]

With data from over one million people, the study found that individuals who engaged in 150 minutes (2 hours and 30 minutes per week, or approximately 20 minutes per day) of moderate-intensity exercise per week had a *28 percent lower risk of death from any cause.*

But the data gets even more compelling.

Those who engaged in 750 minutes per week—or about 1 hour and 45 minutes per day—had a mind-blowing *42 percent lower risk of death* compared to those who never exercised.

> *There is no known therapy or drug with this type of well-studied effect on human lifespan.*

Regular, frequent movement—especially outside, in fresh air—has the potential to transform your mood, maintain your heart health (protecting you from the number one killer in the world!), and significantly boost your creativity and problem-solving skills. That's because when you exercise, your muscles release myokines into your bloodstream. These myokines go to your brain, regulating physiological and metabolic responses and affecting cognition, mood, and emotions. They also aid in the formation of new neurons and increase brain plasticity.

Another recent study found that participants who walked over 4,000 steps a day had healthier brain tissue, better memory, and superior cognitive function compared to those who walked fewer than 4,000 steps per day.[33]

So, if you don't yet exercise regularly, start *somewhere*. Ideally, you'll strive to get 7,000 steps a day plus weight training three times per week. As you can see, just 4,000 steps a day can be the difference between life and death, between a healthy brain and Alzheimer's—or a number of other debilitating diseases.

Benchmarking Your Progress: A Target to Shoot For!

In 2022, a friend, Regan Archibald, collaborated with my coach, Dan Sullivan, founder of Strategic Coach®, to create "Your Fitness 50 Benchmark" as a simple means of self-assessment, goal-setting, and tracking progress.

> **"Living to 100 isn't the problem most of us will face; the problem is the condition in which we arrive."**

Imagine having the fitness levels of the healthiest 50-year-olds at the age of 100. Your future self will appreciate the time you spent taking care of your health today, and the Fitness 50 Benchmarks are a simplified way to measure that investment.

Each of the Fitness 50 Benchmarks can be viewed as a diagnostic predictor of the diseases caused by aging and is an early indicator for all-cause mortality. The Fitness 50 Benchmarks are the easiest way to measure your progress toward your life extension goals—no blood draws, stool tests, or DNA swabs.

The three levels are Fit, Fitter, and Fittest. I see them as motivation to make improvements regardless of my age or current fitness level. I like to work on each of the 11 benchmarks over the course of a week, but you could choose to perform some of the benchmarks in groups or test them as a single activity. The purpose is to have a baseline fitness threshold that you can enjoy when you reach 100.

EXERCISE BLUEPRINT

Below are my June 2023 and June 2024 numbers and a simple chart I will use to follow my progress over the years ahead. Consider creating some benchmarks for yourself and gamifying your exercise program.

Your Fitness 50 Benchmarks

Successfully go to Fit to Fitter to Fittest in each category

FIT		FITTER		FITTEST	
Push Ups: 50 Seconds	15	Push Ups: 50 Seconds	25	Push Ups: 50 Seconds	40
Plank Hold	1 Minute	Plank Hold	2 Minute	Plank Hold	3 Minute
Grip Test: Dead hang	30 Seconds	Grip Test: Dead hang	60 Seconds	Grip Test: Dead hang	90 Seconds
Squats: 50 Seconds	20	Squats: 50 Seconds	35	Squats: 50 Seconds	50
Wall Squat Hold	30 Seconds	Wall Squat Hold	60 Seconds	Wall Squat Hold	90 Seconds
Lunges: 50 Seconds	20	Lunges: 50 Seconds	35	Lunges: 50 Seconds	50
Sit-To-Stand: 50 Seconds	6	Sit-To-Stand: 50 Seconds	8	Sit-To-Stand: 50 Seconds	10
Sit-To-Rise: No Hands	0	Sit-To-Rise: No Hands	1	Sit-To-Rise: No Hands	2
Single Leg Balance	20 Seconds	Single Leg Balance	35 Seconds	Single Leg Balance	50 Seconds
1-Mile Walk/Run/Elliptical	12 Minutes	1-Mile Walk/Run/Elliptical	10 Minutes	1-Mile Walk/Run/Elliptical	8 Minutes
25 Burpees	4 Minutes	25 Burpees	3 Minutes	25 Burpees	2 Minutes
ACHIEVED		**ACHIEVED**		**ACHIEVED**	

Dan Sullivan and Regan Archibald Lifetime Extender Collaboration ©eastwesthealth2022

Peter's Fitness 50 Benchmark Numbers

	Reference Fittest	Peter's June '23	Peter's June '24	Peter's June '25
Push-ups: 50 Seconds	40	**90**	**94**	
Plank Hold:	3 min	**3 min**	**3 min**	
Grip Tests: Dead Hang	90 Sec	**135 Sec**	**130 Sec**	
Squats: 50 Seconds	50	**50**	**50**	
Wall Squat Hold:	90 Sec	**180 Sec**	**200 Sec**	
Lunges: 50 Seconds	50	**41**	**45**	
Sit-to-Stand: 50 Seconds	10	**13**	**15**	
Sit-to-Rise: No Hands	0	**0**	**0**	
Single Leg Balance: 50 Secs.	50 sec	**>120 sec**	**>120 sec**	
1-Mile Walk/Run/Elliptical	8 min	**10 min**	**10 min**	
25 Burpees	2 min	**1.5 min**	**1.5 min**	

An Explanation of Each Category

The following is an explanation of the Fitness 50 categories.

Push-ups: Being able to perform 40 push-ups (modifications are fine) can reduce stroke and heart disease by 96 percent, increase strength, and improve growth hormone.[34]

Plank Hold: Improves posture, strengthens the core, and increases flexibility and endurance. Planks increase basal metabolic rate, decrease fat percentage, and have been shown to improve immunocyte counts.[35]

Grip Strength: An increase in strength reduces blood pressure and risk of stroke, improves testosterone and growth hormone, and signals muscle readiness for the entire body through the brain.[36] It can be a stand-in measurement for full body strength and muscle mass.

Squats: Enhances lower body stability and strength while reducing resting blood pressure and improving lumbar stability.[37]

Lunges: Benefits the posterior muscles (hamstrings, glutes, and lower back) better than other movements. Corrects instability and improves balance.[38]

Sit-to-Stand: Improves core strength, endurance, and balance and is a predictor of mortality.[39]

Sit-to-Rise: Musculoskeletal fitness predictor for longevity, balance, and coordination.[40]

Single Leg Balance: Decreases the chances of strokes and improves cerebral circulation and vascular health.[41] Promotes flexibility and prevents cognitive decline while warding off premature mortality.

1-Mile Run/Elliptical/Rowing/Swimming: Improves cardiovascular function, decreases chances of stroke, and improves endurance and strength. Increases cognitive function and fitness.[42] Rowing, swimming, and the elliptical are just examples of alternatives to running. Feel free to test yourself on any endurance-related activity for 8–12 minutes to see how you fare.

Burpees: Increase blood flow, lower blood pressure, and improve strength, coordination, and endurance. They also stimulate fat burning and metabolism and lower the risk of all-cause mortality.[43]

Chapter 3
Mastering Sleep: Simple Practices That Work

"Sleep is Mother Nature's best effort yet to counter death."
—Matthew Walker, PhD

Getting enough sleep is one of the most underappreciated and most important elements of extending your healthspan. A great night of sleep rejuvenates the body, boosts cognitive function, and powers your immune system. It's the foundation for your health and longevity.

> *"You need seven to eight hours of sleep each night.* If you think you're one of those people who can get away with five or six hours of sleep, the scientific evidence is not on your side."

Sleep expert Dr. Matt Walker, author of the excellent book *Why We Sleep*, says that sleep is the single most effective thing we can do to reset our mental and physical health each day.

As Matt told me during my Abundance Platinum Longevity Trip, "Sleep is Mother Nature's best effort yet at immortality."

There is a direct relationship between how well you sleep and how long you live.

In this chapter, I'll discuss the power of sleep and what I'm doing to improve my sleep and, thereby, increase my healthspan.

The Power of Sleep

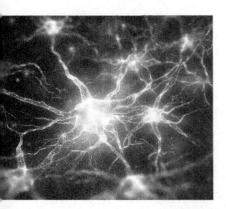

In *Why We Sleep*, Dr. Walker points out that *close to 0 percent* of the total population can get away with less than six hours of sleep per night without harming their health.

For most people, regularly getting eight hours of sleep boosts memory retention, enhances concentration, augments creativity, stabilizes emotions, strengthens the immune system, enhances athletic performance, and staves off deadly ailments like cancers and heart disease.

Still not convinced? Here are three powerful examples:

1. The difference between getting a good night's sleep and a bad night's sleep decreases the brain's ability to retain new facts from 100 percent

to 60 percent—or, as Matt puts it, a sliding scale between "acing an exam and failing it miserably."[44]

2. Going twenty-four hours without sleep is like having a blood alcohol concentration of 0.10 percent, above the legal limit for driving in most places.[45]

3. A sleepless night with only four hours of sleep resulted in a 70 percent reduction in the activity of your natural killer cells. Remarkably, the activity of natural killer cells returns to baseline levels after one night of normal sleep.[46]

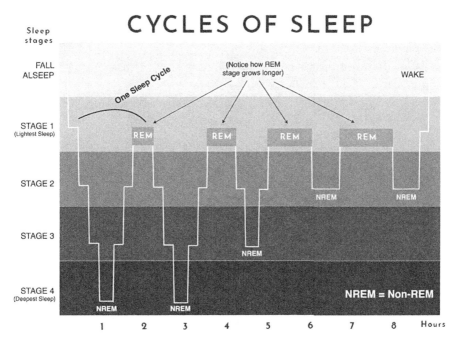

We move through cycles of light sleep, deep sleep and REM sleep every night. (Nationalworld.com)

Sleep plays a critical role in learning and memory. Throughout the night, we fluctuate between light sleep, rapid eye movement (REM) sleep, where dreaming occurs, and deeper non-rapid eye movement (NREM) cycles. These cycles are responsible for transferring the information accumulated throughout the day from short-term memory to long-term memory.

REM sleep and the mental images created during it also help fuel our creativity by generating new connections between ideas that may not be obvious while awake. Dr. Walker's research has shown that REM sleep acts like *overnight therapy* by allowing us to process difficult experiences with rehearsed ease.

Sleep gives our body a vital opportunity to recover from the day's stresses. It aids recovery from inflammation, allows our cells to restock their energy stores, and stimulates muscle repair. During sleep, the glymphatic system in our brains cleans out all the accumulated junk, helping prevent neurodegenerative diseases and keeping our brains working optimally.

Need more evidence? Every spring, when the United States moves the clocks forward one hour for daylight saving time, and we all lose an hour of sleep, hospitals report a 24 percent spike in heart-attack visits around the US.[47] Just a coincidence? Probably not. "That's how fragile and susceptible your body is to even just one hour of lost sleep," says Dr. Walker. "Daylight saving time is a kind of global experiment we perform twice a year," he continues, "and the results show just how sensitive our bodies are to the whims of changing schedules."

With that in mind, here's what I do to ensure I consistently get a good night's sleep.

My Sleep Practices

Forty years ago (during college and medical school), I would pride myself on making do with as little sleep as possible. My target was typically five and a half hours. My usual excuse was, "I'll have plenty of time to sleep when I'm dead." I routinely took red-eye flights to sleep on the flight and hit the ground running. *Boy, was I wrong. I wish I knew then what I know now!*

I now prioritize getting eight hours of high-quality, restful sleep, and here's how I do it.

How Long and When: My target is eight hours of sleep each night, with seven as a minimum. It doesn't mean I always achieve that, but I always try. While I used to be a night owl, routinely staying up until 2:00 a.m., I've shifted my sleep schedule to a much earlier bedtime over the past decade. I'm typically in bed by 9:15 or 9:30 p.m., asleep before 10:00 p.m., and typically wake on my own around 5:30 or 6:00 a.m.

Wind Down and Bedtime: One of the most important keys to getting a good night's sleep is the consistency of your bedtime, complemented by a consistent "wind-down" period.

- **Standard Bedtime:** This is more important than you might expect. Setting and sticking to a routine is critical for high-quality sleep. Eight hours of sleep between 10:00 p.m. and 6:00 a.m. is *not* the same as eight hours between 12:00 a.m. and 8:00 a.m. Consistency is key.

- **Wind-Down Period:** Don't expect to easily go to sleep after a heated Zoom call or an email writing session. Your brain and body need a wind-down period of at least 30 minutes (some say 60 minutes). For me, that wind-down period typically takes place between 9:00 and 9:30 p.m. when I turn down the lights, wear my blue-light-blocking glasses, and slow down my routine. I avoid TV, my phone, and my computer during this period. My last 15 minutes are either a period of meditation (focusing on what I'm most grateful for from that day) or listening to a book on Audible, where I set the countdown timer to 15 minutes. Other wind-down activities include a hot bath (see below), prayer, or reading in bed.

Eye Mask: There are many benefits of wearing an eye mask, and frankly, I have a difficult time sleeping without mine. First, full darkness stimulates higher melatonin levels, helping you fall asleep faster.[48] Second, light exposure disrupts REM sleep, which is essential for cognitive function. Third, it's an easy solution for generating consistent sleeping environments for those who travel a lot (like I do).

I use a Manta Sleep Mask, which I love. It blocks out 100 percent of all light, and the design is super comfortable, avoiding any pressure directly on the eyelid by using a circumferential foam ring around each eye. I've become addicted to my Manta mask and own three, and I always travel with one wherever I go. I'm sure there are other brands, but this is my go-to solution.

Staying Cool at Night: Key research from multiple studies (NIH, American Academy of Sleep Medicine, Harvard) suggests the optimal room temperature for sleep is between 60–67 degrees Fahrenheit (15–19 degrees

Peter's 'sleep-kit' includes: a Manta Mask, Oura Ring, Mandibular Advancement Device, & Blue-light blocking glasses.

Celsius).[49] While the body's core temperature naturally drops in the evening, signaling the onset of sleep, a cooler room temperature can help facilitate the process. Lower temperatures stimulate melatonin production and enhance sleep stages, particularly slow-wave sleep (SWS) and REM sleep. Finally, a cooler room helps prevent overheating, which reduces the likelihood of disruptions like sweating and restlessness. In my routine, there are two things that I do in this area.

- **Room Temperature**: First, I follow the research and set my room air conditioning at a chilly ~63 degrees F (17.2 degrees C), which helps me get my core body temperature down to enter deep sleep.
- **Eight Sleep Cooling Mattress**: Second, I purchased an Eight Sleep, a system that covers the mattress (under the bedding) and cools you down to a chosen temperature at every stage of the night. For example, at the start of the night, I typically set the temp at "-5 degrees," so I get into bed under cool blankets. The nice thing about the Eight Sleep mattress is that you can dial a temperature profile throughout every stage of the night. This means that in the middle of the night, the system can automatically get cooler, and in the morning, I have it set to warm up (+4 degrees), which makes it incredibly pleasant. You can also set the two sides of the bed to operate independently in case your partner likes it set to different temps. Finally, the system also monitors your sleep metrics, like an Oura Ring, to help you track your total, REM, and deep sleep.

Peter loves his Eight Sleep for maintaining perfect sleep temperature profiles and monitoring his sleep through the night.

- **Taking a Hot Bath**: While this isn't part of my personal routine, studies have shown that taking a hot bath 1–2 hours before bedtime can shorten sleep onset latency.[50] It does this in four ways: First, a hot bath raises the body's core temperature. After the bath, the body experiences a rapid cooling process. This drop in temperature mimics the natural decline in core body temperature that occurs before sleep, signaling to the body that it's time to sleep. Second, the warmth of a hot bath can relax muscles and relieve tension, promoting physical relaxation. This relaxation can help reduce stress and anxiety, making it easier to fall asleep. Third, a hot bath increases blood circulation, particularly to the extremities (hands and feet). Improved circulation can enhance the feeling of warmth and comfort, facilitating sleep onset. Finally, the warm water of a hot bath can trigger the release of endorphins, the body's natural "feel-good" chemicals. This release can create a sense of well-being and relaxation, aiding the transition to sleep.

Blue-Light-Blocking Glasses

Our brains evolved in the savannas of Africa to wake up with the sunrise and sleep when the sun sets. Bright blue light, which you'll get from an early sunrise, signals your brain to wake up. Unfortunately, you'll also get bright blue light from your computer screen. Red light, similar to the spectrum of a sunset, is a signal for you to release melatonin and prepare for sleep. In our normal high-tech world, bright lights, computers, cell phones, and TVs just before bed are sleep disruptors. Putting on a pair of blue-light-blocking glasses can help prevent giving your brain the wrong signals.

Given this knowledge, about one hour before going to sleep, I try to put on a pair of blue-light-blocking glasses to ensure my body produces optimal melatonin levels to help me fall asleep more quickly and stay asleep throughout the night. I use a product called TrueDark, which was created and is promoted by biohacker Dave Asprey. On the other side of sleep, when I wake up in the morning, I love to get outside to see the sunrise, to give my visual system and brain the signals to wake up and enter my day fully powered up.

Mandibular Advancement Device

Sleep apnea sucks. It's a sleep disorder characterized by repeated interruptions in breathing during the night, leading to fragmented sleep and decreased oxygen intake. Over time, untreated sleep apnea can contribute to daytime fatigue, impaired cognitive function, mood disturbances, and an increased risk of accidents.[51] Furthermore, it is associated with hypertension, heart disease, stroke, and type 2 diabetes. And, of course, there's a very real association between obstructive sleep apnea and snoring. While snoring occurs due to the vibration of tissues in the throat while breathing in sleep, obstructive sleep apnea involves repeated episodes of partial or complete airway blockage during sleep. These blockages can cause loud snoring, interrupted breathing, and gasping for air, which are hallmark signs of sleep apnea. Therefore, persistent, loud snoring is often an indicator of sleep apnea and warrants medical evaluation. You can test yourself in a sleep lab (there are many you can find just by Googling), or you can try an app like *SnoreLab* to track signs like loud snoring and gasping for air.

In the past, I used a CPAP (continuous positive airway pressure) machine—which is somewhat of a cross between a torture device and the creature from the movie *Alien*. While CPAPs work, they are clumsy and uncomfortable. Now, I use a specially fitted upper and lower mouth guard called a "mandibular advancement device." This device is a dental appliance used to treat sleep apnea and snoring. It is designed to help keep the airway open during sleep by repositioning the lower jaw (mandible) and tongue slightly forward. Plus, it also prevents me from grinding my teeth. I love it so much that I can't sleep without it. There's no single brand I can recommend. You'll need to visit your dentist to get upper and lower impressions taken so the manufacturer can create a perfectly fitted set for your teeth.

Evening Entertainment

As mentioned above, one change I've made during the past five years is to eliminate watching TV in bed before sleep and during my wind-down period. It's well documented that the blue light from screens affects melatonin production. Instead, as discussed, I listen to an audiobook—I guess that's the adult equivalent of being read a bedtime story.

Sleep Supplements and a Pad of Paper

So, what do I do when I have jetlag or trouble getting to sleep? One product I use is Peak Rest by Lifeforce, which includes a slow-release melatonin formulation, magnesium glycinate, L-glycine, L-theanine, and vitamin A. A second hack for me is to have a pad of paper and pen next to my bed to write down my ideas when my brain is overactive in the middle of the night. Getting things out of my head and onto paper allows me to relax and get back to sleep.

Oura Ring: Measuring My Sleep

While the Oura Ring isn't a magical device that will put me to sleep, it does allow me to gamify my sleep by giving me a detailed numeric evaluation of how well I slept through the night. Each morning, one of the first things I do is look at my daily "Readiness Score" and "Sleep Score" (see details at the end of this chapter). My goal is to always get at least a score of 90 on each metric (which I don't always achieve, but it's my target). Often, just the thought that I will be measured in the morning is motivation enough to get to sleep early and minimize any alcohol or late-night food intake.

The other benefits of the Oura Ring include the many other parameters it measures, such as REM, deep sleep, HRV balance, body temperature, and average heart rate.

No Late-Night Eating

One important recommendation for a great night's sleep is to avoid eating within two hours of going to sleep (so, typically, 7:30 p.m. for my desired 9:30 p.m. bedtime). This gives my body enough time to begin digestion and to prevent a full stomach that can lead to regurgitation when prone and subsequent heartburn. Not eating before bed also promotes autophagy, which is how your body cleans out all the misfolded proteins and cellular debris.

The Oura Ring tracks 20+ biometrics.

Your microbiome is also much healthier when you don't eat right before bed since it allows the beneficial organisms that do best in a calorie-poor environment to thrive.

No Caffeine After 12:00 p.m.

It should seem obvious, but any caffeine (coffee, Coke, green tea) in the afternoon and evening can seriously disrupt your desire for sleep later that night. Most of us don't realize that caffeine has a half-life of four to six hours in individuals who are fast caffeine metabolizers and eight to ten hours in slow caffeine metabolizers.[52] This means it takes that amount of time for the quantity of caffeine in your body to be reduced by half. For me, as a slow caffeine metabolizer, having a hit of caffeine with dinner around 6:00 p.m. means I'm effectively drinking half a cup at 10:00 p.m. and then trying to sleep.

Getting Morning Sunlight

There's nothing more effective for jump-starting your day, locking in your circadian rhythm, and boosting your mood than getting out to watch the sunrise. And if you can't wake up that early, at least spend 20–30 minutes outside soaking up natural sunlight as early as possible.

Exposure to morning light enhances wellbeing.

Exposure to natural light in the morning helps synchronize the circadian rhythm, signaling the body to reduce melatonin production, making you feel more awake and alert during the day. Morning sunlight exposure also boosts the production of serotonin, a neurotransmitter that influences mood, energy levels, and focus. Higher serotonin levels during the day are associated with better mood and alertness, and they also help in the synthesis of melatonin at night.[53]

Finally, morning sunlight helps regulate cortisol, often referred to as the "stress hormone." Proper cortisol regulation is essential for maintaining energy levels during the day and ensuring restful sleep at night.[54]

The bottom line is that getting adequate morning sunlight helps you fall asleep faster and experience better quality sleep, and it boosts your mood.

> "Human beings are the only species that deliberately deprive themselves of sleep for no apparent gain. Many people walk through their lives in an under-slept state, not realizing it."
> —Matthew Walker, PhD

Why Sleep Matters

According to a study in *Rand Health Quarterly*, poor-quality sleep costs the US over *$400 billion per year* in lost productivity.[55]

The same study estimates that more than half a million days of full-time work are lost every year due to people sleeping less than six hours.

On an individual level, for most people, sleep is often the first thing they give up. However, the popular belief that "you can sleep when you're dead" is fundamentally damaging to your health, happiness, and longevity.

For example, regularly getting less than six or seven hours of sleep each night doubles your risk of cancer and can increase the likelihood that you'll develop Alzheimer's disease.[56] Insufficient sleep can also contribute to major psychiatric conditions, such as anxiety and depression.[57]

One of the key lessons I took from Dr. Walker's book *Why We Sleep* is the realization that if humans had been able to evolve with the ability to get along with less sleep, we would have. We are most vulnerable to predation and least productive while we sleep. Yet, evolutionarily, our bodies retained the need for eight hours.

Please make getting a solid eight hours of sleep a priority for yourself.

Insights from Matt Walker, PhD: Why Sleep Matters

At my 2024 Longevity Platinum Trip, Matt Walker, PhD, neuroscience and psychology professor and director of the Center for Human Sleep Science at UC Berkeley, shared his insights on how sleep affects our health and longevity. Walker emphasized that insufficient sleep can have serious metabolic and hormonal consequences. He explained that when individuals are sleep-deprived, the body essentially enters a state of malnutrition. "If you take the mentality of 'I'll sleep when I'm dead,' ironically, you will have both a shorter life and the quality of that life will be significantly worse as a consequence." He then highlighted a study where young, healthy men were restricted to five hours of sleep for five nights. Their testosterone levels dropped to the equivalent of someone ten years older. "Five hours a night for five nights will age a man by a decade," Walker emphasized, noting that this also affects female reproductive hormones, such as estrogen and progesterone.

Matt Walker, PhD – Peter's sleep guru.

Walker further explained that sleep deprivation leads to cognitive and metabolic impairments. A study that restricted participants to four hours of sleep for four nights found that these individuals, previously with normal blood sugar, became classified as pre-diabetic by the end of the trial. This underscores the profound metabolic toll that insufficient sleep can take in a very short time.

Deep sleep, Walker noted, "is where the body undergoes its most critical recovery. It acts as a *cardiovascular reboot,* lowering blood pressure and rejuvenating the immune system *by restocking the weaponry in your immune arsenal.*" He also explained how deep sleep helps cleanse the brain by removing harmful proteins like beta-amyloid and tau, which are linked to Alzheimer's disease. For women, Walker pointed out that up to age 51 or 52, they often require more sleep and typically have better deep sleep capability than men. However, after age 60, women's deep sleep capacity sharply declines, which could be linked to menopause—a connection that still needs further research.

Your Checklist for Optimizing Sleep

1. **Take the QQRT Sleep Test**: Visit WhyWeSleep.org to take the free QQRT sleep assessment, evaluating quantity, quality, regularity, and timing.
2. **Aim for 7–9 Hours of Sleep**: Anything less than seven hours can lead to significant metabolic, cognitive, and cardiovascular impairments.
3. **Improve Sleep Efficiency**: Walker emphasizes that efficiency is crucial. Anything less than 85 percent (time spent in bed sleeping versus awake) needs attention.
4. **Maintain Regular Sleep Patterns**: Regularity is more important than quantity when it comes to long-term health benefits. Walker recommended going to bed and waking up at the same time every day, with only a +/- 25-minute variance.
5. **Align Your Sleep with Your Chronotype**: Everyone has a unique chronotype—morning vs. evening person. Aligning your sleep schedule with your natural chronotype (which changes over time) is critical for restorative sleep.
6. **When Not to Exercise for Optimal Sleep**: To maximize your sleep quality, it's best to finish exercising 90 minutes before bed, allowing your body enough time to cool down.
7. **Timing Your Meals for Better Sleep**: It's best to finish eating at least two hours before bed, giving your body time to digest and reduce indigestion during sleep.
8. **Manage Stress Before Sleep**: To avoid "tired and wired" syndrome before bed, consider journaling, taking a hot shower or bath, meditation, and/or speaking to a good friend.
9. **Practice Good Sleep Hygiene**: One hour before bedtime, reduce room lighting by 70 percent. If you need to use your phone, do it *only while standing up*. Put it away when you lie down. Remove clocks from view to avoid watching the time pass.
10. **Monitor Yourself for Sleep Apnea**: Sleep apnea is when breathing repeatedly stops and starts during sleep. It reduces oxygen levels and can lead to increased risks of heart disease, diabetes, and cognitive decline. Use the *SnoreLab* app to track signs like loud snoring or gasping for air. If symptoms appear, seek professional evaluation.

My Oura Ring Sleep Readouts

If you use an Oura Ring, this type of readout (Sleep and Readiness score) on your smartphone will look familiar. My typical Oura score will range between 82 and 97 (my all-time high). My goal every night is to get a sleep score >90. I provide these below as a reference. In a healthy sleep structure, most of your deep sleep should occur in the first half of your night and REM sleep in the second half.

SLEEP SCORE **READINESS SCORE**

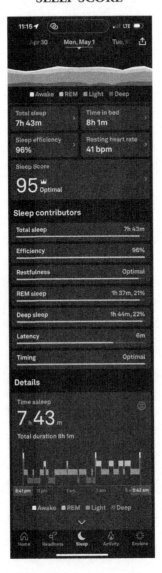
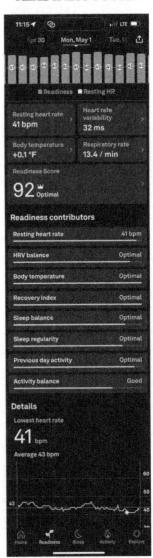

Chapter 4

Don't Die from Something Stupid: Breakthroughs That Can Save Your Life

*"Don't die from something stupid.
The best way to know what is going
on inside your body is to look!"*
—Peter H. Diamandis, MD

Reality: Your Body is Masterful at Hiding Disease

Most of us have *no idea* what's really going on in our bodies. We're all optimists, thinking our health is fine and everything is in great working order. But as it turns out, your body can be quite masterful at hiding the early and middle stages of disease. What do I mean? Consider the following:

- Seventy percent of all heart attacks have no precedent[58]—no pain, no shortness of breath, and, in many instances, nothing to show on a typical coronary CT.[59] Many times, the first symptom of cardiovascular disease will be a heart attack (which can sometimes be fatal). Coronary heart disease affects about 18.2 million Americans age 20 and older, and it kills nearly 366,000 annually.
- Seventy percent of all cancers that are fatal turn out to be the result of cancers that are not routinely tested for by today's medical system (for example, pancreatic cancer or brain cancer like glioblastomas as opposed to colon and breast cancer, which are tested for routinely with colonoscopies and mammograms respectively).[60]
- Parkinson's patients don't develop tremors until nearly 70 percent of the Substantia Nigra neurons are gone.[61] These are the neurons that provide dopamine and control voluntary movements.
- Brain changes in Alzheimer's disease begin years before symptoms appear.[62]
- Most cancer patients have no symptoms from their growing cancer until it has reached stage 3 or stage 4, at which point the chances of a full remission are greatly reduced.[63]

Perhaps you've known a friend or colleague who died in their sleep or someone who went to the hospital with pain only to find out they had some advanced stage of a disease. Here's the issue: Whatever disease was discovered during that medical visit didn't just start that morning; it probably had been going on for some time, perhaps months or years. *Most people have no idea they have a serious medical condition because they don't bother to look.* Most of us know very little about the inner workings of our bodies. We know more about what's going on in our car or refrigerator!

The good news is that your body is also masterful at healing and regenerating itself if you give it a chance. As it turns out, you *can know* with great confidence if you have early cancer, heart disease, an aneurysm, or metabolic or neurocognitive disease. Today, with advanced diagnostics, you can evaluate your health regularly to find diseases at the earliest stage possible and take action accordingly. These advanced diagnostics can also help assess what you're at risk of developing in the future.

This is why I co-founded Fountain Life with Tony Robbins and Dr. William Kapp (Fountain's CEO). Fountain Life is a global platform that provides advanced diagnostics and vetted therapeutics to promote a longer, healthier, and more vital life.

To be clear, Fountain is not the only provider of such services, and you should decide what fits your wallet, geography, and desires. However, you should take action to understand your health status in depth. Many different diagnostic companies offer a wide range of services at varying costs, from $1,000 per year to $150,000 per year.

Some, like Life Force, Viome, InsideTracker, and Function Health, are based solely on blood draws (Viome also incorporates saliva and stool) and don't offer imaging or ongoing concierge medical care as part of the membership.

Others, like Prenuvo, only offer MRI imaging but don't offer CT imaging, genome, blood biomarkers, or concierge care. Yet others, like Human Longevity Inc., Biograph, and Princeton Longevity Center, offer a mix of diagnostic testing similar to Fountain Life.

Fountain Life has multiple locations in the US and is expanding internationally. Current locations in Orlando, Naples, Dallas, and New York. Additional locations open in 2025 in Los Angeles, Phoenix, and Houston.

I will primarily refer to the services offered at Fountain Life since I know them the best and believe they offer the most comprehensive assessment, ongoing care, and therapeutics, making them the leader in proactive, preventative, and personalized longevity.

When I tell someone they can understand in detail what's going on inside their body, some respond that they don't want to know—that they'd rather not find out. For them, my response is:

"You're going to find out eventually. Would you rather find out now, when you can do something about it, or later when it's too late?"

Knowledge is power; almost all chronic diseases can be slowed, stopped, and reversed with the tools and treatments available today. And the first step to regaining your health is knowing what is happening inside your body. Your greatest enemy is ignorance.

My Annual Upload

Every year, I go through a Fountain Life "Upload" (as part of their APEX Membership), which gathers more than 200GB of data about me.

> **"My Fountain Upload answers two key questions: Is there anything going on inside my body that I need to know about? And what is likely to happen to me, and how do I optimize my health to avoid future problems?"**

The upload focuses on finding any of the leading causes of shortened health and lifespan, specifically cardiovascular disease, cancer, aneurysms, metabolic disease, and neurocognitive dysfunction. It also identifies underlying dysfunctions that can, over time, result in disease. When these dysfunctions, such as mitochondrial dysfunction and microbiome dysbiosis, are identified and addressed, diseases can be prevented, and if already present, they can be reversed in many cases.

What I love about Fountain Life is that a decision was made early on to base the entire team on a functional medicine, or root cause, approach. Should any dysfunction be discovered, a functional medicine-trained team will employ a

Fountain Life offers preventative, predictive, & personalized care.

systems-based approach to uncover the molecular drivers and provide an ongoing plan to restore molecular, cellular, and physiological health.

Fountain Life is designed for individuals who want to optimize their lifestyle based on the most advanced comprehensive understanding of what's happening inside their body, their unique risks, and their potential. Members have access to world-class, functional-medicine, longevity-trained physicians who interpret their test results from 200 GB of data from their upload (in partnership with AI), recommend the optimal course of action, and provide treatment as needed. Every member is also assigned a dedicated team that includes a health coach, nutritionist, medical assistant, nurse, and care coordinator for the entire year. The team's goal is to work with our members to build routines and make lifestyle choices conducive to longevity, as well as keep them on track.

A summary of the wide range of testing offered to members is shown on the next page.

Fountain's Advanced Diagnostic Testing Summary

As you research options to meet your needs, below is an overview of Fountain's APEX diagnostic testing for comparison purposes.

Imaging

- **Brain MRI and MRA** (Magnetic Resonance Angiogram with AI): Looks for brain lesions, volumetric changes, aneurysms, and even perivascular spaces that are early warning signs your brain is not efficiently cleaning out debris, which happens during sleep.
- **Full Body MRI**: Looks for cancers, organ abnormalities, aneurysms, liver fat, liver iron, infections, and other soft tissue abnormalities.
- **Coronary CCTA w/ AI:** Measures and characterizes the plaque in your coronary arteries, which is especially important for identifying vulnerable soft plaque that is more prone to rupture.
- **High-resolution 3D Breast Ultrasound:** Perfect for dense breasts and implants.
- **Low-Dose Lung CT**: Looks for cancer, nodules, fibrosis, and infections.

- **DEXA Scan**: Measures bone density, muscle mass, fat mass, and inflammatory visceral fat.
- **Retinal Scan with AI**
- **Photometric Dermatology Screening with AI**

FUNCTION TESTING

- Balance Testing
- Grip Strength Testing
- Olfactory Testing (an early sign of cognitive dysfunction)
- Pulse Wave Velocity (indicates peripheral artery stiffness)
- Electrocardiogram (EKG)
- Oral Microbiome
- Gut Microbiome Assessment

Blood Testing

- Whole Genome Sequencing
- Blood-Based Multi-Cancer Early Detection Testing
- Comprehensive Blood Biomarkers (100+)

Getting quarterly blood biomarker testing.

- Lipids (Basic lipids, Lp(a), ApoB, omegas)
- Metabolics (Comprehensive metabolic panels, HbA1c, Insulin, Homocysteine)
- Inflammation and Oxidation Markers (Fibrinogen, Hs-CRP, TMAO)
- Hormone Panel
- Thyroid Panel
- Hematology: CBC
- Nutrients CoQ10, B12, Fe, Mg, K, Se, Zn

Optional Blood Testing (a la carte testing beyond baseline APEX membership)

As you work with your longevity physician, it may be necessary to dive deeper into certain areas, such as nutritional testing, heavy metals, mold, environmental toxins, and neuro testing. Your physician can arrange for additional testing on an a la carte basis. Your team will order, interpret, and incorporate the results of these tests into your plan.

Why Should You Look?

As I mentioned, the body is amazing at compensating for any disease and, thereby, masking its impact. Additionally, we are all optimists about our health. You think you're doing fine until that moment when you're rushed to the hospital.

The numbers tell the story. Out of 100 seemingly healthy adults who underwent a Fountain Life screening, here are our initial findings.

- 2.0 percent had a cancer they didn't know about.
- 2.5 percent had an aneurysm.
- 14.4 percent of seemingly healthy members had a serious finding they needed to address immediately (e.g., cardiovascular, metabolic, or neurocognitive disease).
- Almost all members had findings that should be addressed to prevent disease and optimize their health.

Given that heart disease is the number one killer for both men and women globally—accounting for one out of every four deaths in 2021—here are a few additional (shocking) data points.[64] Note that these findings are for adults who have received the best healthcare available in the conventional medical system and believe they are in good or excellent health when they enter the Fountain ecosystem.

- 76 percent of Fountain members were found to have stage 1 coronary artery disease.
- 15 percent of members were found to have stage 2 or 3 coronary artery disease.

- 35 percent of members with stage 1 disease had only *soft plaque* (the most dangerous type) that would have been missed on a traditional "calcium scan."

The good news is that, given the advances in modern medicine, there are now therapeutics (drugs, supplements, foods, and lifestyles) that can be used to prevent and, in most cases, reverse heart disease, metabolic disease, and, in some cases, even dementia. Again, your greatest advantage in fighting disease is early detection, which gives you the greatest chance of taking successful action.

I call going through my annual Fountain Life upload "not dying from something stupid." I feel a great sense of confidence and relief when I complete each one.

By the way, also under the category of "not dying from something stupid," I include wearing your seatbelt, not texting while driving, and wearing a helmet when you ski.

More Details on Diagnostic Testing: What and Why

The following is a more detailed outline of my annual multi-modal testing. Fountain's AI system and my functional medicine-trained physician and medical team organize and evaluate all data.

It's worth pointing out that blood tests alone and imaging alone are never really sufficient. Each modality can miss something on its own. Fountain Life prides itself on taking a "multi-modal" approach that combines dozens of tests, thereby increasing the probability of discovering and validating disease early. Following are two examples within the Fountain Life data set from early 2024.

1. Prostate cancer was found on an MRI with a normal PSA and normal blood-based multi-cancer early detection test (such as GRAIL).
2. Head and neck cancer was found on a blood test with an initial normal MRI read.

How frequently should someone conduct a multi-modal "upload"? For me, this is an annual visit (lasting about 4-5 hours), and frankly, I wouldn't dare skip a year. It's important to understand that both heart disease and cancer can develop and progress significantly within a single year. Your upload should not be considered a "one-and-done" protocol. Again, knowledge is power.

Imaging

- **Full Body and Brain MRI with AI Overlay:** MRI imaging is used to generate a quantitative 3D visualization of your body, looking for tumors, masses, aneurysms, and other anomalies in the brain, head, neck, chest, kidneys, liver, spleen, pancreas, and other abdominal organs, as well as the bladder, uterus, prostate, and other pelvic organs. The imaging also measures all brain structures, looking for white-matter lesions and shrinkage. MRIs are very safe and do not utilize radiation.

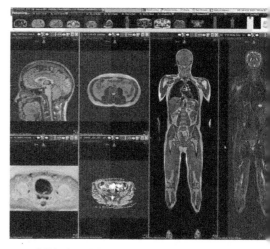

MRI full-body image can identify aneurysms, cancers, and anomalies in the brain and body.

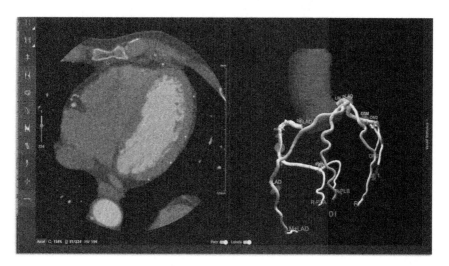

A Coronary CT with a CLEERLY AI Overlay enables you to find your soft plaque, which is the most dangerous type to address.

- **CCTA with AI Overlay:** Coronary computed tomography angiography (CCTA) is used to evaluate coronary artery disease. CCTAs use radiation and typically deliver a radiation dose in the range of 2–10 millisieverts (mSv). For comparison, on average, people are exposed to a background radiation dose of about 2–3 mSv per year from natural sources (such as cosmic radiation and radon), and a single chest X-ray typically exposes a person to about 0.1 mSv of radiation. Fountain uses a special type of AI overlay on their CCTAs called CLEERLY that quantifies the location and type of plaque in the coronary arteries. As it turns out, calcified plaque, measured as a "calcium score," is not concerning unless it significantly blocks (occludes) a coronary artery. Calcified plaque is stable; think of cement inside an artery. The most concerning and dangerous is "soft plaque," a lipid-rich plaque with a soft covering that can more easily rupture. When this happens, a clot will form immediately around the ruptured plaque, blocking the coronary artery, stopping blood flow, and depriving the heart muscle of oxygen, which is what we call a heart attack. The CLEERLY AI-overlay scan is designed to identify and measure all types of plaque, particularly soft plaque.

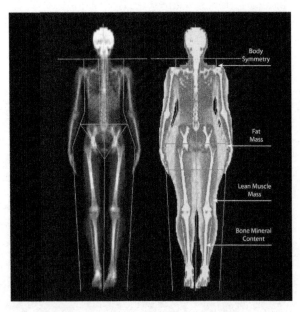

DEXA Scan image of bones, muscle, and visceral fat.

- **DEXA Whole Body Scan:** The DEXA, or "dual-energy x-ray absorptiometry" scan, is used to measure bone density (osteopenia and/or osteoporosis) and body composition (muscle and fat). It is critical to get a DEXA scan if you are over 50, especially if you are female and have certain risk factors, such as having a parent who has experienced a broken hip.[65] However, our initial data shows that even young men have osteopenia (thinning of the bones). Quantifying muscle mass is critical to determine if you have sarcopenia or low muscle mass since muscle is the organ of longevity. The amount of fat and type of fat you have in your body is also critical to monitor since it can put you at risk for metabolic diseases such as type 2 diabetes. It turns out that visceral fat, the fat around your organs, is very inflammatory and a big risk factor for all chronic diseases.[66] The amount of radiation used in DEXA scans is very low, typically around one-tenth of the dose used in a standard x-ray.

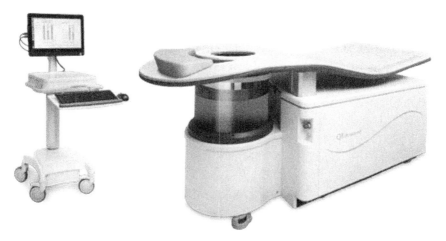

Quantitative Transmission Ultrasound used to 3D image breast tissue.

- **QT (Breast) Imaging:** QT imaging, or quantitative transmission ultrasound imaging, is a breast imaging technique that uses low-frequency sound waves to create 3D images of breast tissue. It can be used to help detect and diagnose breast cancer and is especially useful for dense breasts, young women, and women with breast implants. QT imaging offers high-resolution images of the breast and has several advantages over traditional mammography, including no radiation, no injections, and no uncomfortable breast compression.

- **Skin Cancer Screening:** This screening is provided to help catch skin cancer early. According to the American Cancer Society's estimates, approximately 97,610 new cases of melanoma will be diagnosed in the United States in 2023 (about 58,120 in men and 39,490 in women).[67] About 7,990 people are expected to die of melanoma (about 5,420 men and 2,570 women). Fountain Life uses an innovative photo-based AI system that a dermatologist then confirms.

- **Retinal Scan:** The retina at the back of the eye is the only place in the body where you can directly visualize blood vessels. Diseases like diabetes, high blood pressure, and even high cholesterol can all show up as changes in the retina. The camera captures high-resolution images of the retina that AI reads to rule out the presence of diabetic retinopathy.

Genomics

Two decades ago, the human genome was first sequenced. Since then, science has been able to decipher more and more of what is encoded in the 3.2 billion nucleotides you received from your mother and father. Fountain Life offers whole genome sequencing, which identifies the sequence of almost 100 percent of your DNA, generating a data file 150 GB in size (part of our 200 GB Upload). In comparison, other genetic tests such as 23andMe only look at SNPs (single nucleotide polymorphism) in your genome, reading as little as 0.02 percent of your DNA.[68]

Blood Biomarkers, Cell-Free DNA, and Toxins

Following are the extensive blood tests available to help evaluate your health status.

- **100+ Blood Biomarker Tests**: These tests contain an extensive and comprehensive set of clinical biomarkers looking at kidney and liver function; vitamin, mineral, and hormone levels; vascular and full body inflammatory markers; complete lipid and cholesterol panels; blood cell characterization; extensive metabolic health markers; autoimmune disease screening markers; cell membrane analysis; and a full set of heart disease risk biomarkers.
- **Multi-Cancer Early Detection Blood Test:** Did you know you can detect free-floating cancer cells in your body with a simple blood test? The Galleri test identifies over 50 types of cancer, including hard-to-detect ones like ovarian, pancreatic, liver, and esophageal

cancers. It analyzes circulating tumor DNA in the blood—free-floating DNA shed by cancer cells. The test uses advanced technology to examine DNA methylation patterns, which differ between healthy and cancerous cells. This allows for early detection of cancer signals and the prediction of where in the body the cancer originated, offering a powerful, minimally invasive tool for early cancer detection.

- **Testing for Toxins**: Fountain Life screens for a variety of environmental and biological toxins that can contribute to acute or chronic health issues. These include endocrine-disrupting chemicals (EDCs), heavy metals, microplastics, and other harmful substances that can accumulate in the body. An example of these toxins are PFAS, or per- and poly-fluoroalkyl substances, which are a group of man-made chemicals in our cookware, clothing, and food packaging that do not break down and are known as "forever chemicals," causing serious health issues, including cancer, liver damage, immune system impairment, and developmental effects in infants and children.

Metabolic, Mitochondrial, and Neurocognitive Testing: These tests are available to APEX members a la carte as additional diagnostic tools.

- **Continuous Glucose Monitoring (CGM)**: As mentioned earlier, I love wearing my CGM to give me real-time glucose readings day and night that provide insight into what foods and drinks spike my blood glucose levels.

- **Nutritional Test (Genova NutrEval)**: A blood and urine profile that evaluates over 125 biomarkers and assesses 40 antioxidants, vitamins, minerals, essential fatty acids, amino acids, and digestive support. This test also evaluates all of the key metabolites in mitochondrial function.

- **Mitochondrial Function Test**: Testing lactic acid generated by a brief workout (treadmill or stationary bike) to determine your Zone 2 heart rate, estimate your VO2, and detect the presence of any mitochondrial dysfunction.

- **Neurocognitive Test**: A series of computerized neuropsychological tests to evaluate neurocognitive status. These cover a range of mental processes from simple motor performance, attention, and memory to executive functions and screen for mental health and mood disorders.

Other Testing

- **Microbiome Sequencing**: Full sequencing of your gut microbiome, coupled with an AI interpretation that ingests all the published literature on the microbiome to understand how it affects your overall health and what can be done to move it to a healthier pattern. The microbiome has been implicated in every aspect of human health and function, especially the gut, brain, and immune system.
- **Oral Pathogen Test**: Quantitative test for the most critical bacterial pathogens that confer the highest risk for infection, inflammation, cardiovascular, immune, and cognitive diseases.
- **Grip Strength Test**: Grip strength is a key indicator of health and is correlated to a range of health issues, including cardiovascular disease, cancer, osteoporosis, diabetes, immune function, mental health, and overall mortality.

Cost and Future Offerings

I want to acknowledge that such extensive testing and year-round concierge medical care is not cheap and will likely be unaffordable by many reading this book. Today, an APEX Membership has an *all-inclusive* membership price of $19,500. The annual membership includes all the testing, bloodwork, imaging, and an AI-powered app. It also includes a full year of concierge medical coverage (which includes a functional medicine concierge MD, a nurse practitioner, a health coach, and a dietitian). The good news is that, as with all things "technology," volume and exponential improvements bring down costs over time.

As of the end of 2024, at the time of publication, Fountain Life is rolling out a membership called CORE, which is offered to corporations for a price of $6,500 per employee for companies that sign up for ten or more memberships. While this includes all the imaging, genomics, and some blood biomarker testing, it only offers a single review of results with a functional medicine physician and year-round access to the AI-powered app. It does not include access to the concierge functional medicine team on an ongoing basis. If this is of interest, consider discussing Fountain CORE with your employer.

Why This Matters

In 2004, my dear friend Ray Kurzweil wrote a milestone book called *Fantastic Voyage: Live Long Enough to Live Forever.*

In the book, Kurzweil describes three different "bridges" to get from now through to longevity escape velocity. *Bridge one interventions* include a set of near-term life-extending solutions, such as those described in this book (e.g., a glucose-minimized diet and muscle-mass-increasing exercise), that should be followed until we get to bridge two. *Bridge-two interventions* include technologies like cellular reprogramming, synthetic organs, and stem cell therapies, which we expect to be fully tested and routinely available in the clinical setting later this decade. This gets us finally to what Kurzweil describes as the third bridge. *Bridge-three interventions,* expected in the mid to late 2030s, involve the impact of nanotechnology when nanobots (human or AI-designed molecular machines) are available to repair your biology on an atomic and molecular basis.

Your job is to stay healthy and free of accidents long enough to intercept many of the bridge-two therapies expected later this decade.

Helping you cross "bridge one" is this book's goal and Fountain Life's mission.

If this interests you, and I hope it does, check it out at FountainLife.com. In the appendix of this book, I've also shared URL links to other diagnostic companies discussed at the beginning of this chapter.

Fountain Life's Massive Transformative Purpose (MTP):

"Our MTP is to extend member healthspan using the most advanced diagnostics and vetted personalized therapeutics."

Some Final Motivation

In July 2024, I joined Raoul Pal for the second time on his *Real Vision* podcast to speak about longevity and AI, and I mentioned our work at Fountain Life.

The day the podcast came out, I was sincerely moved by a number of the comments made by members of his community. If this were an exception, I would still be super proud, but we have hundreds of similar examples of saving people's lives.

No matter where you get your multi-modal testing done, it is critical that you take responsibility for your own health and look carefully at what's happening inside your body.

GO — **George OMalley** — Aug 1, 2024 | 1:39 PM

The original interview you did with Peter led me to FountainLife. Saved my life- found i had 90% blockage in multiple areas, and had triple bypass surgery. My cardiologist who was not affiliated with FountainLife said me going to take the test saved my life

---

EB — **Eddie Burke** — Aug 1, 2024 | 4:26 PM

WOW - what a great discussion - your previous discussion with Peter encouraged me to pay for a local body scan - they found an aneurysm - I had no idea - this is currently being monitored to make sure that it is not growing and is not in the danger zone for size - you guys are great - thanks

---

AG — **Anahata Graceland** — Aug 3, 2024 | 1:55 PM

I went to fountain Life, Peters company with Tony Robins that helps to create a total review of your health, body, vascular system, bones, organs genetics biome and more.... Thanks to the AI and their good work, I had a diagnosis of a very serious thing and now can get it addressed. AI is remarkable and a gift from those who were impressed to create with it. Great Interview.

Chapter 5

Your Longevity Pharmacy: Medications, Supplements, & Cutting-Edge Therapies

"The medicines that will allow us to live to 150 years or more with a high quality of life are in clinical trials today."
—Greg Bailey, MD

Peter's Legal Disclosure

I am an educator, entrepreneur, and scientist. While I have a medical degree, I am not a licensed and practicing clinician and cannot make clinical recommendations for the prevention or treatment of any disease. In providing the details below, I am sharing the list of supplements, medications, and therapeutics I'm using based on my physician's recommendation.

No one should start taking any supplements or medications without first checking with their personal physician.

Some supplements can be dangerous for people with certain pre-existing medical conditions, and supplements can interfere with some prescription drugs. Supplements can also affect different people differently. While the FDA reviews supplements to determine if their listed ingredients are safe to consume, no US regulatory authority has reviewed a supplement's ability to address specific medical conditions. The evidence of benefit for most supplements comes from laboratory experiments and/or from epidemiology data, not from human clinical trials. Supplements should only be purchased from trusted retailers and brands. Testing has shown that many supplements are tainted with unlisted ingredients and/or do not contain the amount of the supplement listed on their label.

Chapter Overview: Navigating My 75 Meds & Supplements

This chapter is somewhat difficult to write because medications and supplements need to be personalized for every individual. Everyone is dealing with different genetics, lifestyle choices, and age-related health issues, which, in combination, should determine exactly what you take. Your medications and supplements need to be selected in consultation with your physician based on your goals, blood biomarkers, and the number of pills you're willing to take each day.

Having said that, I will present the logic and rationale for the 75+ different medications and supplements I take every day.

This chapter is about the 75 pills I take every day, what they are, and why I take them.

It might seem extreme, but every day (morning, lunch, and evening), I take *over 75* different medications and supplements as part of my daily regimen to boost longevity and healthspan. You're probably wondering why anyone would take so many pills. How do you select what to take, and do they work?

In this chapter, I'll divide the discussion into three key parts: (1) my prescription medications, (2) my supplements, and (3) the therapeutic treatments I'm exploring.

My Prescription Medications

Note: The prescription medications listed below are specific to my medical objectives and based on my blood biomarkers (which I test roughly every quarter) and the recommendations of my Fountain Life physicians. You will need to determine what is appropriate for you in consultation with your physician. My goal is only to fully disclose my current regimen for informational purposes.

My Cholesterol Medications

In the United States, heart disease has been the leading cause of death since 1950, accounting for 702,880 deaths in 2022. While the rate of heart disease in men is higher than in women, it is also the leading cause of death for women, killing one in five women each year in the US. It can affect men and women at any age, and about 80 percent of women between the ages of 40 and 60 have at least one risk factor for it. Heart disease is a broad term that includes many conditions, such as coronary artery disease, angina, heart failure, and arrhythmias. Risk factors for heart disease include high blood pressure, high cholesterol, smoking, diabetes, obesity, unhealthy diet, physical inactivity, and excessive alcohol use.

> **"In the United States, heart disease has been the leading cause of death for both men and women since 1950, accounting for 702,880 deaths in 2022."**

One of the risk factors based on my family history and genetics is heart disease secondary to hypercholesterolemia. As a result, I'm keenly following numerous factors to monitor my cholesterol status, such as my LDL, HDL, ApoB, and Lp(a) levels. In addition, I track my blood pressure regularly (checking it occasionally when I wake or at night before going to sleep). I also depend on the CLEERLY AI overlay with my CCTA cardiac imaging to look for soft plaque (as described in Chapter 4).

While I monitor and care about my LDL, HDL, and triglyceride levels, there is increasing evidence that *heart disease may be even more correlated to Hemoglobin A1c* (HbA1c) levels—a measure of your average blood glucose (sugar) over the past 2 to 3 months—than with any other blood biomarkers.[69] HbA1c reflects the percentage of hemoglobin (a protein in red blood cells that carries oxygen) that is glycated (has sugar attached to it). A higher HbA1c level indicates higher blood glucose levels over time.

Elevated blood sugar levels can cause inflammation in blood vessels, making them more susceptible to atherosclerosis (plaque build-up).[70] Elevated levels also damage the endothelium (the lining of blood vessels), leading to increased stiffness and reduced elasticity of the arteries, which can contribute to high blood pressure, increasing the risk of heart attacks and strokes. As discussed in the "Longevity Diet" chapter, sugar is a poison, and minimizing it in my diet is one of my key objectives.

While high cholesterol is often associated with an increased risk of heart disease, low cholesterol levels can also pose health risks in a few key areas. Cholesterol is a precursor to steroid hormones, including cortisol, estrogen, and testosterone. Very low cholesterol levels can lead to a decrease in these hormones, potentially causing hormonal imbalances that affect mood, energy levels, and overall health.[71] In addition, cholesterol is vital for brain function; it's a key component of brain cells, especially the protective coating of neurons called myelin, and is involved in the synthesis of neurotransmitters.[72] Low cholesterol levels have been associated with a higher risk of depression, anxiety, and cognitive decline.[73] Finally, cholesterol is important for absorbing fat-soluble vitamins (A, D, E, and K). Low cholesterol levels can impair the absorption of these essential nutrients, leading to deficiencies and associated health problems.

When tuning your medications and supplements, it's important to have specific target numbers. For someone who has not had a heart attack or other heart condition, here are the recommended guidelines:

- Cholesterol <200
- LDL <100
- HDL >40 for men >50 for women
- Triglycerides <100
- Apo B <90

Measuring cholesterol precursors, as is done at Fountain Life, can help determine what type of medication may be most effective for you. For example, if you absorb most of your cholesterol rather than making it, taking a medication that inhibits this absorption will be much more effective.

I personally use a number of medications to maintain my ideal ranges.

- **Zetia** (10 mg): This medication treats high levels of cholesterol. Unlike other cholesterol-lowering agents, ezetimibe selectively inhibits the intestinal absorption of cholesterol and related phytosterols. This medication is taken orally daily.
- **Crestor** (5 mg): I take Crestor (rosuvastatin) at the lowest dose. I use this statin for reasons beyond its impact on lowering LDL (bad cholesterol) and triglycerides. It also has anti-inflammatory effects (established in various acute and chronic inflammatory models) as well as antiviral and antioxidant properties. It also helps stabilize any vulnerable plaque. Other benefits of Crestor are that it is

hydrophilic and doesn't cross the blood-brain barrier; therefore, it may not affect neurocognition.
- **Repatha** (PCSK9 biologic): This biologic medication (a monoclonal antibody) lowers LDL cholesterol levels. It works by blocking a protein called PCSK9, which inhibits the liver's ability to remove LDL cholesterol from the blood. This is injected (with an autoinjector) every two weeks.

Rapamycin: King of Longevity Medications

There is one medication that, in the opinion of many experts in the field, is likely to be the most impactful longevity therapeutic available at this moment. It's called rapamycin. Discovered in the soils of Easter Island (Rapanui) and originally used as an immunosuppressant, this compound now stars in the lineup of longevity enhancers for its fascinating properties. It inhibits mTOR, a crucial protein that governs cell growth and proliferation. But for those of us chasing the fountain of youth, rapamycin's real appeal might lie in its ability to tweak our cellular cleanup processes and metabolism. (**Note:** The clinical trials in humans are just beginning to prove this.)

Every Sunday night, my ritual includes a 6 mg dose of rapamycin (taken as three 2 mg tablets) taken once per week for three months, followed by one month off, and then repeat. The goal is to modulate my immune system to support my body's age-fighting mechanisms. Rapamycin was originally taken daily by organ transplant patients and used to suppress the immune system to prevent the human body from rejecting transplanted organs. There are a multitude of longevity-related medications and supplements (which I discuss below), and out of all of them, rapamycin is the one longevity-related intervention in my arsenal that seems to have the most concrete evidence backing its benefits.

Here's some of the data: In 2009, *Nature* published a groundbreaking study showing rapamycin increased mouse lifespans by 14 percent for females and 9 percent for males.[74] Fast forward to 2014, and *Science* reported that it enhanced immune responses in humans, prompting a 20 percent greater response to flu vaccines in elderly individuals.[75] This finding shattered the misconception that rapamycin weakened the immune system when taken intermittently.

Dr. Matt Kaeberlein, a professor at the University of Washington, has been researching rapamycin for 20 years and found that a three-month course of rapamycin increased remaining life expectancy in middle-aged lab mice by up to 60 percent. As he puts it:

> *"I would say that rapamycin is the current best-in-class for a longevity drug that we have."*

Moving beyond mice, more recently, a 2023 *Geroscience* study surveyed 504 adults, revealing that 65.5 percent of the 333 people taking rapamycin believed in its anti-aging effects.[76] Nearly half reported improved health, with over 35 percent saying their brain "works better" and 38 percent feeling younger.

Several human clinical trials are currently exploring the potential of rapamycin as a longevity medicine, and they are showing promising early results. One notable trial is the Participatory Evaluation (of) Aging (with) Rapamycin (for) Longevity Study (PEARL). This large-scale, placebo-controlled study is being conducted to assess the effects of rapamycin on aging in humans. The PEARL study involves 200 participants aged 50 and older who will receive rapamycin for up to one year.

My Rapamycin Dosing: 0.1 mg/kg once weekly (which for me is 6 mg) to be cycled for three months on and one month off. TruAge testing before and six months after.

Note: Before you consider taking rapamycin, it's crucial to consult with a qualified healthcare provider who can assess your health, risk factors, and potential benefits. Longevity research is still in its early stages, and there are many unknowns and risks associated with using drugs like rapamycin for this purpose. Here are the known downside complications: In larger dosages, Rapamycin suppresses the immune system, which can make individuals more susceptible to infections. Rapamycin can affect glucose metabolism and insulin sensitivity, potentially leading to insulin resistance and an increased risk of type 2 diabetes. Rapamycin use can cause side effects such as diarrhea, nausea, and vomiting, which may reduce the quality of life for some individuals. Long-term use of immunosuppressive drugs like rapamycin has been associated with an increased risk of certain types of cancer, particularly skin cancer and lymphoma. Rapamycin can be hard on the kidneys, potentially leading to impaired kidney function, which can be problematic, especially for older individuals who may already have reduced kidney function. Rapamycin can interact with other medications and may have different effects on different individuals.

CJC/Ipamorelin: Peptide to Boost Growth Hormone and IGF-1

One culprit that may cause age-related changes, such as sarcopenia, osteopenia, reduced energy levels, and cognitive function, is the natural fall-off in human growth hormone (HGH) produced by our pituitary glands.

Beyond the role played by HGH during childhood and adolescence in promoting growth and development, HGH continues to provide a critical function well into adulthood. The benefits of adequate HGH levels include maintaining muscle mass and strength, preserving adequate bone density, promoting fat metabolism, maintaining skin elasticity, and supporting wound healing and sleep quality.

The challenge is that HGH production is estimated to decrease by about 14 percent every decade after age 30. This means that by age 60, HGH levels may be significantly lower than in early adulthood.

While some have chosen to supplement HGH directly by injecting the hormone, numerous studies suggest that long-term use of HGH may increase the risk of certain cancers, such as breast and colon cancer. And other studies show that HGH can increase the risk of heart problems, including heart enlargement and heart failure.

An alternative I've chosen to use is the combination of two peptides: CJC-1295 and Ipamorelin (CJC/Ipamorelin). CJC-1295 is a growth hormone-releasing hormone (GHRH) analog that stimulates the pituitary gland to *produce* HGH, and Ipamorelin is a growth hormone-releasing peptide (GHRP) that directly stimulates the *release* of HGH from the pituitary gland. These substances are known as secretagogues since they stimulate the secretion of HGH.

There are many potential benefits of CJC/Ipamorelin. Following is a list of all of the anti-aging attributes thought possible using these peptides secondary to an increase in HGH and GHRH:

- Increased muscle mass and strength
- Improved bone density
- Faster wound healing
- Enhanced fat burning
- Improved sleep quality
- Increased energy levels
- Improved cognitive function

It's also been proposed that CJC/Ipamorelin may help reduce the visible signs of aging, such as wrinkles and sagging skin, and improve cardiovascular health by lowering cholesterol levels and improving heart health.

Dosage and Administration: CJC/Ipamorelin is typically administered by subcutaneous injection once or twice per day. The peptide is injected under the skin, usually in the abdomen, thigh, or back of the upper arm. Dosage and administration can vary depending on individual factors, such as age, weight, and health conditions. The typical dosage ranges from 200–600 micrograms per day, as advised by your physician.

Note: CJC/Ipamorelin is a prescription medication and should only be used under the guidance of a healthcare professional. Also, as of the writing of this book, the FDA is conducting an extensive evaluation of peptides and may restrict the sale of this class of therapeutics.

Testosterone Supplementation and Optimization

If you're a man, an important question is whether or not you should supplement your natural testosterone levels. (See the women's chapter for a discussion on female hormone replacement.) I've chosen to do this based on my quarterly blood test results for the following reasons. This next section is intended to give you context to evaluate for yourself (in consultation with your physician) what is best for you.

The first point is to realize that our testosterone levels change over time. As we age, we may be faced with low testosterone levels, known as hypogonadism, and supplementation may have numerous benefits for longevity and overall health.[77] To understand why these changes occur, it's valuable to remember that from an evolutionary standpoint (and I'm speaking about our ancestors hundreds of thousands of years ago), our bodies were never designed to live past age 30—the point at which we had reproduced, and our progeny had also reproduced. After that point, our physiology had no reason to maintain the same levels of hormones, such as testosterone and growth hormones, found in our twenties. During prehistoric times, before the development of language, after the age of 30, we were a burden to our clan, eating the food that might otherwise go to our grandchildren; therefore, the best thing we could do to help perpetuate our species was to die. It's brutal but true.

> **"From an evolutionary standpoint, *our bodies were never designed to live past age 30*—the point at which we had reproduced, and our progeny had also reproduced."**

For that reason, it's not surprising to see how our testosterone levels change over time, steadily dropping over the decades.

Ages 18–30: Testosterone levels are at their peak during late adolescence and early adulthood. During this period, men experience high levels of testosterone, which contribute to the development of secondary sexual characteristics, peak muscle mass, bone density, and sexual function.

Ages 30–40: After reaching peak levels, testosterone declines gradually, typically about 1 percent per year.

Ages 40–60: This decline becomes more noticeable during middle age, resulting in reduced libido, fatigue, decreased muscle mass, and potential changes in mood and cognitive function. This period is referred to as "andropause" or "male menopause."

Ages 60 and Beyond: By this stage, significantly lower testosterone levels may lead to further reductions in muscle mass, increased body fat, decreased bone density, osteoporosis, increased fatigue, depression, and diminished sexual function.

It is for all of the above reasons that some men choose to supplement their endogenous testosterone, known as testosterone replacement therapy (TRT) or what I prefer to call "testosterone optimization."

If you are considering testosterone optimization (in coordination with testing and your physician's advice), the following are some of the benefits that are typically expected.[78]

- **Heart Function**: Improved heart function, especially in men with heart failure.
- **Blood Lipids**: Helps regulate cholesterol levels, potentially lowering LDL (bad cholesterol) and increasing HDL (good cholesterol), reducing the risk of heart disease.
- **Muscle Preservation**: Testosterone supplementation can help maintain or increase muscle mass and strength, preventing frailty and sarcopenia.

- **Metabolic Rate:** Increased muscle mass can boost metabolic rate, helping prevent obesity, which is associated with numerous age-related diseases.

- **Bone Health:** Testosterone is essential for maintaining bone density. Supplementation can help prevent osteoporosis, reducing the risk of fractures.

- **Cognitive Health:** Some studies suggest that testosterone supplementation may help maintain or improve cognitive function, including memory, attention, and processing speed, which can decline with age.

- **Mood Enhancement:** Testosterone may also have positive effects on mood, reducing symptoms of depression and anxiety, which can impact overall quality of life.

- **Libido and Sexual Function:** Low testosterone is associated with reduced libido and erectile dysfunction. Supplementation can improve sexual health.

- **Blood Sugar Regulation:** Testosterone supplementation may improve insulin sensitivity, reducing the risk of type 2 diabetes. Better blood sugar control is associated with a lower risk of many age-related diseases.

- **Body Composition:** Testosterone helps reduce fat mass, particularly visceral fat, which is linked to a higher risk of metabolic syndrome, cardiovascular disease, and other age-related conditions.

- **Inflammation Control:** Chronic inflammation is a key driver of aging and age-related diseases. Testosterone supplementation may have anti-inflammatory effects, which could contribute to increased longevity.

Some experts suggest that maintaining total testosterone in the young, healthy range of 700–900 nanograms per deciliter (ng/dL) is reasonable and healthy.[79] It's important to also check free testosterone, sex hormone binding globulin (known as SHBG), dihydrotestosterone (DHT), and estradiol to look at the whole pattern. Insulin, cortisol, and thyroid levels, as well as IGF-1, are important to monitor. There is a delicate interplay between all these elements. It is also important to remember that testosterone acts by binding to the testosterone or androgen receptor, and there is significant genetic variability between different people that affects how strongly testosterone will bind to this receptor. Therefore, testosterone levels need to be optimized for the individual and how they feel rather than just the number.

Based on my lab levels, I supplement with a 0.15ml injection once per week. My goal is to keep my total testosterone in the range of 800–900 ng/dL. I check my testosterone levels every quarter to ensure I am within my desired optimal range. *Do not supplement without testing.*

Modafinil: My Nootropic Medication of Choice

I often fly around the world, and as much as I love getting eight hours of sleep, I frequently need to go straight into a meeting or jump onstage after crossing the continent or flying overseas. In those cases, I utilize a nootropic medication to focus my mind and get a boost of alertness.

Nootropics, often called "smart drugs" or "cognitive enhancers," aim to improve cognitive function, particularly executive functions like memory, creativity, or motivation in healthy individuals.

Caffeine is by far the most common nootropic used worldwide. Because I'm a slow metabolizer of caffeine, I've reduced my intake to one cup of caffeinated coffee per day, and then I switch to decaf. Beyond caffeine, other nootropics include amphetamines such as Ritalin and Adderall. These stimulants increase levels of dopamine and epinephrine in the brain. Dopamine is a neurotransmitter known to regulate attention span, movement, and pleasure, while norepinephrine is a stimulant that increases energy. I choose not to use these for many reasons, including their high potential for abuse and dependency, as well as side effects like anxiety, cardiovascular issues, and insomnia.

Another common nootropic is nicotine, which can be taken orally as a nicotine gum or lozenge (available in 2 mg and 4 mg doses). When we think of nicotine, we typically think of cigarettes and lung cancer. The reality is that cigarettes cause cancer primarily because of the carcinogenic chemicals in tobacco smoke. These chemicals include tar, benzene, formaldehyde, nitrosamines, and radioactive elements like polonium-210, among others. When tobacco is burned, these substances are inhaled into the lungs, where they can directly damage cells and DNA, leading to cancer.

If you smoke or vape, please quit. It is the number one pro-longevity action you can take—more important than anything else you've read thus far.

Independent of the carcinogenic chemicals in tobacco, nicotine can have a mood-stabilizing effect, providing a temporary boost in mood and reduction in anxiety. I personally avoid using nicotine because it is also highly addictive, leading to physical dependence and withdrawal symptoms, which can include irritability, anxiety, and difficulty concentrating. It also increases heart rate and blood pressure, contributing to an elevated risk of heart disease, stroke, and other cardiovascular problems.

Modafinil: What I have found to work very well *for me* is a prescription medication known as modafinil (or Provigil) (my typical dose is 100 mg). The medication provides me with added alertness, focus, and clarity.

Modafinil is a medication that promotes wakefulness. It's most commonly used to treat sleep disorders such as narcolepsy, daytime sleepiness secondary to obstructive sleep apnea, and shift work sleep disorders. Despite its primary use, modafinil has also gained popularity as a nootropic due to its potential to enhance cognition, particularly by increasing alertness and reducing fatigue. It is used by professionals, such as pilots, who require alertness over long periods.

The exact mechanism of action of modafinil is not fully understood. It affects various neurotransmitters in the brain—chemicals neurons use to communicate with each other. Some key neurotransmitters modafinil increases include dopamine, norepinephrine, serotonin, glutamate, histamine, and orexin (hypocretin). It also suppresses GABA, which is an inhibitory neurotransmitter.

Modafinil is thought to increase the availability of these neurotransmitters in parts of the brain that control wakefulness and alertness. For instance, it's known to inhibit the reuptake of dopamine, which increases the amount of dopamine available in the brain. This is a similar mechanism to some stimulants, but modafinil doesn't tend to cause the same level of overstimulation or potential for addiction.

I've found modafinil beneficial and have used it under medical care with no observed side effects. However, that is me, and the use of modafinil by others is not without risks and potential side effects.

Note on Modafinil Side-Effects: It's important to use modafinil only under the guidance and supervision of a qualified healthcare provider, especially if you have underlying health conditions or are taking other medications. While it is generally considered safe and well-tolerated when used as prescribed, there are some potential downside risks associated with its use, including headache, nausea, nervousness, anxiety, and gastrointestinal issues. Sleep disruption is common if taken too late in the day. Modafinil may cause increases in blood pressure and heart rate in some individuals. People with preexisting cardiovascular conditions should use it cautiously and under medical supervision. While rare, modafinil has been associated with psychiatric side effects such as mood changes, agitation, and, in rare cases, hallucinations. Modafinil should not be used when consuming alcohol. It's also important to remember that modafinil doesn't replace the need for sleep.

My Supplements

Getting back to the 75+ pills I take daily (the majority of which are supplements), questions you might be asking yourself are: What are they? What do they do? How do you choose what to take?

My goal in this section is to provide you with a logical structure addressing what I take and why I take them to help you evaluate what might be right for you. I am always researching, learning, and adjusting my supplement regimen every quarter after my quarterly testing and in consultation with my physician.

The 12 Hallmarks of Aging

The primary construct I use to evaluate and organize my supplements comes from a paper published in 2013 in the prestigious journal *Cell* called the "Hallmarks of Aging."[80] The paper focused on the underlying causes of aging (on a molecular and cellular basis), organizing them into nine major causes or "hallmarks" that need to be addressed if extending healthspan and slowing aging is your goal.

Since 2013, the number of hallmarks has been increased from nine to twelve.

> **My medication and supplement routine is built around addressing as many of the twelve hallmarks of aging as possible.**

The 12 hallmarks of aging are *genomic instability, telomere attrition, epigenetic alterations, loss of proteostasis, deregulated nutrient-sensing, mitochondrial dysfunction, cellular senescence, stem cell exhaustion, altered intercellular communication, disabled macroautophagy, chronic inflammation, and dysbiosis.*

Below are the supplements I take and how they map against the hallmarks of aging. In each case, I'll discuss what I am taking and why but will refrain from providing details on any dosage levels and frequency. Again, this is for you to discuss with your physician and may depend on your diet since many of these naturally occurring compounds are found in our foods.

It's also worth noting that in many cases, you'll see the same supplement appearing over and over again under multiple hallmarks. In this case, I only discuss the supplement once in detail when first mentioned and then simply note its utility when it appears in later hallmarks.

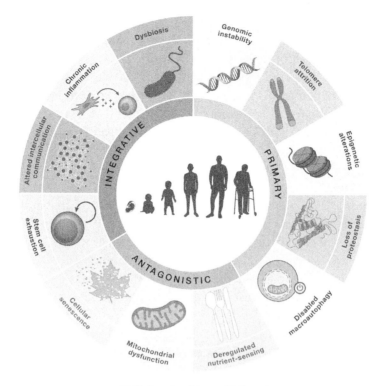

12 Hallmarks of Aging. (Cell.com)

1. Genomic Instability

Genomic instability refers to the increased rate of mutations in the genome that occurs with aging, which can lead to various age-related diseases, including cancer.[81] Several supplements are thought to counteract genomic instability by protecting DNA, improving DNA repair mechanisms, and reducing oxidative stress, which can damage DNA. Here's what I use:

Nicotinamide Mononucleotide (NMN): NMN is a precursor to NAD+, a critical molecule in DNA repair processes. NAD+ activates sirtuins, which are enzymes involved in metabolism, stress response, inflammation, and cell survival, as well as maintaining genomic stability by promoting the repair of DNA damage. The benefits include improved DNA repair and supporting cellular energy metabolism, potentially reducing the risk of accumulating mutations in the DNA. An alternative to NMN is nicotinamide riboside (NR), which is also a precursor to NAD+, and resveratrol, a polyphenol found in red wine. Resveratrol is also known to activate the sirtuin family of proteins, which help in DNA repair and maintenance of genome integrity (see section below on NAD+ boosters).

Coenzyme Q10 (CoQ10) with PQQ: Another protective supplement against genomic instability is CoQ10 and pyrroloquinoline quinone (PQQ), antioxidants that protect cells from oxidative damage. It is a fat-soluble molecule in all biological membranes in our bodies. It helps maintain mitochondrial function and is a component of the enzyme involved in making ATP. CoQ10 also reduces the oxidative burden on DNA and cell membranes because it acts as an antioxidant and is involved in regenerating the important antioxidants: vitamins C and E. Statin medications reduce the body's ability to make CoQ10, so I add CoQ10 for all of the benefits above and because I take Crestor.

Vitamin C is a powerful antioxidant that neutralizes free radicals and reduces oxidative stress, protecting DNA from damage. Humans and guinea pigs are the only mammals unable to make Vitamin C and must get it from outside sources. Vitamin C is a water-soluble vitamin.

Vitamin E is another potent antioxidant that protects cell membranes and DNA from oxidative damage, supporting cellular integrity and reducing the risk of genomic instability. It is a fat-soluble vitamin and is important for protecting the fatty cell membranes. There are many different types of Vitamin E sold as supplements. I prefer to use natural forms rather than synthetic forms, which are made from petroleum products and are less potent. There are eight different natural forms of Vitamin E, so I take mixed tocopherols and tocotrienols since they more closely represent what we obtain from our food.

Folate (Vitamin B9) is essential for DNA synthesis and repair. Folate is involved in the methylation process, which is crucial for genomic stability. I monitor my folate levels throughout the year, as well as my homocysteine, which can increase if I don't get sufficient Vitamin B12 and folate.

Magnesium Threonate and Magnesium Glycinate: Magnesium is a cofactor in over 300 enzymatic reactions, including those involved in DNA repair and replication. Stress readily depletes our body of magnesium, so it is important to supplement magnesium whenever under stress. Magnesium is also needed for muscle relaxation and gut motility. If I ever feel my shoulders tighten or start to become constipated, I know it is a sign that I likely need more magnesium. It is an essential mineral involved in numerous physiological processes. Magnesium threonate can get into the brain and thus has been suggested to enhance brain function and support cognitive health, while magnesium glycinate helps reduce stress, promote muscle relaxation, and maintain a balanced nervous system. It addresses many of the hallmarks of aging, including enhancing mitochondrial function, reducing chronic inflammation, protecting against genomic instability, supporting proteostasis, delaying cellular senescence, improving nutrient sensing, maintaining stem cell function, and promoting neuroprotection.

Curcumin, the active compound in turmeric, has powerful anti-inflammatory and antioxidant properties that can protect DNA from oxidative damage. Curcumin neutralizes free radicals and may stimulate the action of other antioxidants for a synergistic effect. Curcumin has also been shown to boost brain-derived neurotrophic factor (BDNF), which is a critical factor in learning, memory, and overall brain health. Curcumin also protects the function of the inner lining of blood vessels, has been shown to be beneficial in treating cancer, and can help with arthritis, Alzheimer's disease, and even depression. It has such ubiquitous benefits likely because of its powerful anti-inflammatory and antioxidant effects.

Quercetin is a flavonoid with strong antioxidant properties. Flavonoids are phytochemicals found in plants like broccoli, onions, and apples. Quercetin can help protect DNA from oxidative stress by reducing the impact of free radicals on DNA. Quercetin acts as a powerful anti-inflammatory and can reduce the risk of cancer, neurodegenerative diseases, high blood pressure, and heart disease. Quercetin has been shown to stabilize mast cells and can effectively prevent allergic symptoms.

Selenomethionine is a form of selenium, a trace mineral that acts as an antioxidant and supports immune function. Selenium is a trace element that is a component of selenoproteins, which have antioxidant properties and are involved in DNA repair. Selenoproteins also produce hormones, such as thyroid hormone, and help regulate the immune response. Selenium is found in plants that obtain the selenium from the soil in which they are grown. Since many soils are low in selenium, selenium deficiency is becoming more common. It is a critical component of several antioxidant enzymes, including glutathione peroxidases and thioredoxin reductases. These enzymes protect cells from oxidative damage by neutralizing reactive oxygen species (ROS). Selenium is essential for the thyroid gland to function properly, and it is required to synthesize thyroid hormones and convert thyroxine (T4) to the more active triiodothyronine (T3). Adequate selenium levels have also been associated with boosting immune function, reducing inflammation, protecting cardiovascular health, and providing neuroprotection.

2. Telomere Attrition

Telomere attrition, which refers to the shortening of telomeres (the protective caps at the ends of chromosomes), is another hallmark of aging.[82] As telomeres shorten with each cell division, they eventually reach a critical length that triggers cellular senescence or apoptosis (cell death), contributing to aging and age-related diseases. Several supplements are thought to help counteract telomere attrition by promoting telomere maintenance or lengthening.

Astragalus (TA-65) is a traditional herb used in Chinese medicine. Derived from the astragalus root, this molecule is believed to activate telomerase, the enzyme responsible for maintaining and extending telomeres. Telomerase is not active in most cells after embryonic development but is still found in male germ cells, activated lymphocytes, and certain stem cells.

Omega-3 Fatty Acids have anti-inflammatory properties and may protect against telomere shortening by reducing oxidative stress and inflammation. Omega-3 fatty acids act as a precursor to anti-inflammatory molecules and are found in our cell membranes. They are essential for heart, brain, and skin health. Maintaining the correct ratio of omega-6 to omega-3 fatty acids is critical. Most of us get far more omega 6s in our diet than omega 3s, so it is important to test your levels to ensure you have the right amount.

Vitamin D with K2 is thought to play a role in telomere maintenance, potentially through its effects on reducing inflammation and supporting immune function. Vitamin D also plays a crucial role in bone health and immune function. When combined with Vitamin K2, it helps ensure calcium is properly utilized and deposited in bones, potentially reducing the risk of fractures and supporting overall health as we age. Vitamin K2 has also been shown to reduce arterial stiffness by decreasing microcalcifications and lowering the incidence of diabetes and cardiovascular disease.

Zinc is an essential mineral that plays a role in DNA repair and synthesis, including telomere maintenance. Zinc is also an essential micronutrient for immune function and wound healing. It is important for a proper sense of taste and smell, as it helps the cells lining the intestinal tract absorb nutrients and can help lower blood sugar and cholesterol levels. Zinc is also important for the proper activity of thyroid hormone.

Also, slowing telomere attrition is a number of supplements already discussed above. These include **Nicotinamide Mononucleotide (NMN), Nicotinamide Riboside (NR), Resveratrol, Vitamin C, Vitamin E, Folate (Vitamin B9), and CoQ10.** Reducing oxidative stress in the cells also helps maintain telomere length. I should note that mindfulness/meditation has also been shown to positively affect telomeres.

3. Epigenetic Alterations

Epigenetic alterations refer to changes in gene expression that do not involve changes to the underlying DNA sequence.[83] These changes can accumulate over time and contribute to aging and age-related diseases. Supplements that target epigenetic alterations generally work by influencing the enzymes and pathways regulating gene expression, such as DNA methylation, histone modification, and non-coding RNA interactions.

Not surprisingly, many supplements that help battle genomic instability also help with epigenetic alterations. From our listing above, we see a number of repeat superstar supplements, including **Resveratrol, NMN, NR, Curcumin, Quercetin, Folate (Vitamin B9), and Vitamin D.**

New to our supplement list are the following:

Sulforaphane: This powerful compound is found in cruciferous vegetables like broccoli (which, by the way, is one of my favorite foods to eat). Sulforaphane works its magic at the very core of our cells, influencing how our genes are expressed by modulating something called histone deacetylases. Imagine your DNA as a long string of instructions tightly wound around proteins called histones. These histones act like spools, keeping the DNA compact and organized. Sometimes, the DNA must be unwound from these spools to read certain genes. This is where histone deacetylases come in; they're like tiny machines that help keep the DNA tightly wound, potentially silencing some genes. Sulforaphane can modulate the activity of these histone deacetylases, essentially helping to "loosen" or "tighten" the DNA around the histones as needed.

Along with influencing DNA methylation (another way cells control gene expression), this process allows sulforaphane to promote detoxification processes and support the maintenance of healthy gene expression patterns. Sulforaphane also acts at another level to modify which genes are expressed by activating a protein called Nrf2. Nrf2 is a transcription factor that, when activated, enters the nucleus and binds to sites on the DNA called antioxidant response elements (AREs) in the promoter of endogenous antioxidant genes. Essentially, it turns on the expression of our antioxidant defense mechanisms. In simpler terms, sulforaphane helps ensure the right genes are turned on or off at the right times, contributing to overall cellular health and potential longevity.

Green Tea Extract (EGCG): Do you ever wonder why everyone says green tea is so healthy? As it turns out, epigallocatechin gallate (EGCG) in green tea is known to modify DNA methyltransferases and histone deacetylases, enzymes involved in epigenetic modifications. The catechins in green tea, such as EGCG, are a type of flavonoid that also helps fight inflammation and oxidative stress.

S-Adenosyl Methionine (SAMe) is a key methyl donor in the body. SAMe acts as the universal methyl donor and provides the methyl groups that our methyl transferase enzymes use to methylate many different compounds in the body, including our DNA. DNA methylation is a critical epigenetic process and essential for normal gene expression and aging. As we age, the pattern of methyl groups on our DNA changes. We can even use this information to create clocks that measure our biological age.

Alpha-Ketoglutarate (AKG) is a metabolite that influences epigenetic regulation by serving as a cofactor for enzymes that modify histones and DNA. AKG is a molecule that is an intermediate in the Krebs cycle, which is how our cells make energy. It has been shown to delay the onset of age-related decline in model organisms, such as yeast and mice.

4. Loss of Proteostasis

Loss of proteostasis refers to the decline in the body's ability to maintain the proper function, folding, and degradation of proteins.[84] This can lead to the accumulation of damaged or misfolded proteins, which can contribute to age-related diseases like Alzheimer's and Parkinson's. Several supplements are thought to support proteostasis by enhancing protein folding, reducing oxidative stress, and promoting the degradation of damaged proteins.

Not surprisingly, some of the supplements we've seen above are back on the list again. These include **Resveratrol, Curcumin, CoQ10, Omega-3 Fatty Acids, and Flavonoids (such as Green Tea Extract and Quercetin).**

New to our supplement list are the following:

High-Quality Essential Amino Acids, particularly leucine, are important for muscle protein synthesis and repair. Amino acids are the building blocks of our proteins, and without the proper building blocks, we can't synthesize proteins. Our body can make most of the amino acids needed to build our proteins, but there are a number of amino acids that we cannot make and must be obtained from what we eat. These nine amino acids are called essential amino acids. Amino acids are used to make muscle protein and all of our enzymes.

N-Acetylcysteine (NAC) is the rate-limiting precursor to glutathione, one of the body's most important antioxidants and detoxifiers. Glutathione protects proteins from oxidative damage and supports the detoxification of harmful substances, helping maintain proteostasis. If you have been exposed to any type of toxin (which is hard to avoid in our current environment), it is very important to support your glutathione levels by taking NAC.

Ginseng and Ashwagandha (also known as Heat Shock Proteins [HSPs] Inducers). Heat shock proteins are molecular chaperones that assist in protein folding and protect cells from stress. Supplements like ginseng and ashwagandha may induce the production of HSPs, thereby enhancing the cell's ability to manage damaged or misfolded proteins. Ashwagandha is known for its beneficial effects on reducing stress and anxiety, likely by modulating cortisol levels. It has been shown to support a healthy immune response and help maintain good blood sugar and testosterone levels. It has antioxidant activity to protect brain cells and is used to promote overall memory. It has been a traditional medicine for thousands of years. As with many of these ancient herbal remedies, we are now beginning to understand from a scientific perspective why they can be so effective. Maintaining proteostasis is one of those ways.

Alpha-Lipoic Acid (ALA) is a universal potent antioxidant that acts on its own as well as helps recycle other antioxidants, such as glutathione, Vitamin C, and Vitamin E. ALA works in both water and fat-soluble tissues so that it can benefit all parts of our cells in all our tissues. It is used to decrease insulin resistance in our cells and regulates AMPK in the hypothalamus, decreasing feelings of hunger. It can also protect our cells from radiation damage. I take ALA before flying since air travel exposes us to significant levels of radiation. Alpha-lipoic acid is one of those superstar supplements that does not get enough recognition.

Pterostilbene is a compound similar to resveratrol with potentially higher bioavailability. Pterostilbene activates sirtuins and other pathways involved in the stress response and proteostasis. It may also enhance autophagy and the clearance of damaged proteins, supporting proteostasis.

5. Deregulated Nutrient-Sensing

Deregulated nutrient sensing refers to the impaired ability of cells to sense and respond to nutrients, leading to metabolic imbalances that contribute to aging and age-related diseases.[85] The key nutrient-sensing pathways involved in aging include insulin and IGF-1 signaling, mTOR, AMPK, and sirtuins. Supplements that target these pathways can help restore balance and counter the effects of deregulated nutrient sensing.

Again, some supplements we've seen earlier are back on the list. These include **Resveratrol, NMN, NR, Alpha-Lipoic Acid, Omega-3 Fatty Acids, Quercetin, Curcumin, CoQ10, and Flavonoids (e.g., Green Tea Extract).**

New to our supplement list are the following:

Berberine is a plant alkaloid that activates AMP-activated protein kinase (AMPK), a key enzyme involved in energy balance and nutrient sensing. Berberine improves insulin sensitivity, reduces blood glucose levels, and helps modulate lipid metabolism, thereby supporting healthy aging. Berberine has also been shown to have a beneficial effect on the gut microbiome by targeting pathogenic microbes and promoting the growth of beneficial bacteria. Berberine is a very bitter compound, and since bitter taste receptors in the gut activate GLP-1 secretion by L-cells in the intestine, it acts as a natural GLP-1 agonist.

Note: Metformin, a prescription medication, is another widely used diabetes medication that activates AMPK, promoting energy balance and improving insulin sensitivity. It is associated with a reduced risk of age-related diseases and may extend lifespan by mimicking the effects of calorie restriction and improving nutrient sensing. In general, metformin is a very safe and well-tolerated medication. However, it's worth noting that some studies suggest that metformin may inhibit the mTOR pathway, potentially reducing the effectiveness of resistance training and muscle growth.

6. Mitochondrial Dysfunction

Mitochondrial dysfunction refers to the decline in the function and efficiency of mitochondria, which are the powerhouses of our cells.[86] Mitochondria are responsible for producing the majority of the cell's energy through a process known as oxidative phosphorylation, which generates adenosine triphosphate (ATP), the cell's primary energy currency. As we age, mitochondria become less efficient at producing ATP, leading to a decrease in cellular energy levels.

Other elements of mitochondrial dysfunction include:

Impaired mitophagy is the process by which damaged or dysfunctional mitochondria are selectively degraded and removed from the cell. With age, mitophagy becomes less efficient, resulting in the accumulation of defective mitochondria, which contributes to cellular dysfunction and aging.

Reduced mitochondrial biogenesis, or the creation of new mitochondria, is another dysfunction leading to a decrease in the overall number and quality of mitochondria in cells.

Finally, mitochondrial DNA (mtDNA) damage is yet another dysfunction. Since mitochondria have their own DNA, separate from the nuclear DNA, over time, mutations can accumulate in mtDNA due to its proximity to the reactive oxygen species generated during ATP production. These mutations can impair mitochondrial function, leading to further energy deficits and cellular dysfunction.

There are several supplements known to support mitochondrial health and function.

Many of the supplements we've seen earlier are back on the list again. These include **CoQ10, NAD+ Precursors, such as NMN and NR, Resveratrol, Alpha-Lipoic Acid, Magnesium, Curcumin, Omega-3s, and Folate.**

New to our supplement list are the following:

Urolithin A is a compound produced by certain microbes in the gut from ellagitannins found in foods like pomegranates, berries, and nuts. It has recently gained significant attention for its potential anti-aging properties, particularly due to its ability to enhance mitochondrial function. Urolithin A is best known for stimulating mitophagy, addressing mitochondrial dysfunction, and allowing old mitochondria to be selectively degraded and recycled. Urolithin A's role in enhancing mitophagy helps keep mitochondria healthy and efficient. It also has an antioxidant effect and can clear free radicals and promote ATP production, specifically in muscle cells. One trial even showed that Urolithin A could stimulate mitochondrial gene expression in sedentary elderly adults in a manner normally seen in healthy adults who exercise regularly.

Acetyl-L-Carnitine (ALCAR): L-carnitine is a naturally occurring amino acid. The acetyl-L-carnitine form is better absorbed and crosses the blood-brain barrier more rapidly than L-carnitine. LCAR and ALCAR transport fatty acids into the mitochondria, where they are used for energy production. ALCAR also supports mitochondrial function and has neuroprotective effects by supporting mitochondrial function in the brain.

MitoQ: A specialized form of CoQ10 attached to a lipophilic cation, designed to penetrate mitochondria more effectively by crossing the fatty cell and mitochondrial membranes more easily. MitoQ acts as a powerful antioxidant within mitochondria, protecting them from oxidative damage.

7. Cellular Senescence

Cellular senescence is a process where cells stop dividing and enter a state of permanent growth arrest without dying.[87] While senescent cells can play beneficial roles in wound healing and preventing the proliferation of damaged cells, their accumulation over time contributes to aging and age-related diseases. These cells secrete harmful inflammatory factors, known as the senescence-associated secretory phenotype (SASP), which can damage surrounding tissues.

Several supplements have been studied for their potential to counteract cellular senescence by either reducing the burden of senescent cells, modulating the SASP, or supporting overall cellular health.

Back on our list of beneficial supplements are a few superstars mentioned above, including **NAD+ Precursors such as NMN and NR, Vitamin C, EGCG (Epigallocatechin Gallate), Curcumin, Quercetin, Pterostilbene, Berberine, Ashwagandha and Astragalus Membranaceus (TA-65).**

New to our supplement list are the following:

Fisetin is another flavonoid found in fruits like strawberries. It has been shown to act as a senolytic, helping reduce the burden of senescent cells and improve healthspan in preclinical studies. Fisetin may also reduce the SASP, thereby mitigating the harmful effects of senescent cells on surrounding tissues. There is still some controversy about the correct dosing and timing of the dose to achieve this effect.

Dasatinib, while not a supplement per se (dasatinib is a prescription drug), when combined with natural compounds like quercetin, shows potent senolytic activity. This combination has been shown to clear senescent cells from tissues, improving function and reducing inflammation. It is being studied for its potential to promote longevity.

Note: This is not a medication I am currently taking. When I recently measured my cellular senescence using a test called SapereX, I found my levels to be rather low and not in need of any reduction.

8. Stem Cell Exhaustion

Stem cell exhaustion is our eighth hallmark of aging, characterized by a decline in the function and regenerative capacity of stem cells.[88] This reduces tissue repair and maintenance, contributing to age-related decline and diseases. Certain supplements and compounds are thought to help counteract stem cell exhaustion by supporting stem cell health, promoting their regenerative capabilities, and reducing the factors that lead to their exhaustion.

Many of our previous supplement superstars are on the list here again, including **NAD+ precursors NMN and NR, Resveratrol, Fisetin, Curcumin, Omega-3 Fatty Acids (DHA and EPA), Alpha Lipoic Acid, Astragalus Membranaceus (TA-65), Quercetin, Acetyl-L-Carnitine (ALCAR), EGCG (Epigallocatechin Gallate), and Vitamin D.**

New to the supplement list, supporting our stem cells are the following:

Spermidine is a natural polyamine found in certain foods and produced in the gut microbiome. It promotes autophagy, the process of cellular self-cleaning. Spermidine supports the health and function of stem cells by enhancing autophagy, which helps maintain cellular quality control and delay the exhaustion of stem cells. It has also been shown to improve gut barrier integrity, protect against a leaky gut, and support a healthy gut microbiome.

Rhodiola Rosea is an adaptogen known for its stress-reducing properties. Rhodiola may help protect stem cells from stress-induced damage, thereby reducing the rate of stem cell exhaustion.

9. Altered Intercellular Communication

Altered intercellular communication is a hallmark of aging that refers to the breakdown in the way cells communicate with each other.[89] This disruption can lead to chronic inflammation, immune system decline, and other age-related changes in cellular behavior. Addressing this issue involves modulating inflammatory pathways, supporting immune function, and improving the signaling mechanisms between cells.

It's no surprise that many of the supplements previously listed may help counter altered intercellular communications including **NAD+ Precursors (NMN, NR), Omega-3 Fatty Acids (DHA and EPA), Resveratrol, Curcumin, Quercetin, Vitamin D, Fisetin, EGCG (Epigallocatechin Gallate), Spermidine, Alpha Lipoic Acid, Ashwagandha, and Berberine.**

New to our supplement list are:

Probiotics are beneficial bacteria that support gut health and immune function. A healthy gut microbiome is crucial for maintaining proper immune signaling and reducing systemic inflammation, which is key for healthy intercellular communication.

Carnosine is a dipeptide with antioxidant and anti-glycation properties. Carnosine can protect cells from the harmful effects of advanced glycation end-products (AGEs) and oxidative stress, which can impair cellular communication.

Melatonin is a hormone that regulates sleep-wake cycles and has antioxidant properties. Melatonin has anti-inflammatory effects and can improve immune function, which is essential for maintaining proper intercellular communication as we age.

10. Disabled Macroautophagy

Disabled macroautophagy, often referred to simply as autophagy, is the tenth hallmark of aging.[90] Autophagy (from the Greek terms "self" and "eating") is the process by which cells degrade and recycle their components, such as damaged proteins and organelles, to maintain cellular health. When autophagy is impaired, these damaged components accumulate, leading to cellular dysfunction and contributing to the aging process.

As expected, many of the superstar supplements discussed above also support autophagy. They include **NAD+ Precursors (NMN and NR), Spermidine,**

Resveratrol, Curcumin, Berberine, Quercetin, EGCG (Epigallocatechin Gallate), Pterostilbene, Sulforaphane, and Melatonin.

New to the supplement list, supporting autophagy are the following:

Astaxanthin is a carotenoid with potent antioxidant effects derived from microalgae. Astaxanthin can enhance autophagy and reduce oxidative stress, supporting cellular repair and maintenance. It is particularly useful for eye health and can help prevent cataracts and age-related macular degeneration.

Fasting Mimetics: While not a supplement, fasting mimetics, such as time-restricted eating, ProLon (a "fasting-mimicking diet"), and drugs like rapamycin (which can replicate some of the biochemical and physiological effects of fasting, particularly through its inhibition of the mTOR pathway) support autophagy.

11. Chronic Inflammation

Our eleventh hallmark is chronic inflammation, a persistent, low-grade inflammatory state that contributes to the development of various age-related diseases.[91] Addressing chronic inflammation is crucial for promoting healthy aging and preventing the progression of inflammatory-related conditions.

The anti-inflammatory supplements discussed in earlier hallmarks that also address chronic inflammation include the following: **Omega-3 Fatty Acids (such as DHA and EPA), Curcumin, Quercetin, Resveratrol, Vitamin D, Zinc, Magnesium, Green Tea Extract (EGCG), S-adenosylmethionine (SAMe), Alpha-Lipoic Acid, N-Acetylcysteine (NAC), CoQ10, and Probiotics.**

New to the supplement list, helping minimize chronic inflammation, are the following:

Boswellia Serrata (Frankincense) is an herbal extract with potent anti-inflammatory properties. Boswellia inhibits the 5-lipoxygenase enzyme, which is involved in producing pro-inflammatory leukotrienes. This makes it effective for reducing inflammation, particularly in conditions like arthritis.

Ginger is a root with anti-inflammatory and antioxidant properties. Ginger reduces inflammation by inhibiting the production of pro-inflammatory cytokines and prostaglandins. It has been shown to be effective in reducing pain and inflammation in conditions such as osteoarthritis.

12. Dysbiosis

Our final hallmark is Dysbiosis, which refers to an imbalance in the gut microbiome, where harmful bacteria outnumber beneficial ones.[92] This imbalance can contribute to inflammation, weakened immunity, and various age-related diseases, making it a significant concern in the context of aging. Addressing dysbiosis is essential for maintaining gut health and overall well-being.

Here are some supplements that may help counteract dysbiosis:

Probiotics, mentioned earlier, are live beneficial bacteria that help restore and maintain a healthy gut microbiome. Probiotics can help balance the gut flora by introducing beneficial bacteria, reducing the growth of harmful bacteria, and enhancing gut barrier function. Different strains like *Lactobacillus*, *Bifidobacterium*, and *Saccharomyces boulardii* are particularly effective. A significant amount of research is being done on different probiotics, and soon, we will be able to take specific probiotics for specific conditions.

Prebiotics are non-digestible fibers that feed beneficial gut bacteria. Prebiotics like inulin, fructooligosaccharides (FOS), and galactooligosaccharides (GOS) promote the growth of beneficial bacteria in the gut, helping restore balance and counteract dysbiosis. The beneficial bacteria consume these fibers, turning them into short-chain fatty acids like butyrate (see below).

Cal/Mag Butyrate supplements support gut health and overall well-being. Calcium/magnesium butyrate promotes a healthy gut microbiome, enhancing digestion and supporting immune function, thus contributing to a balanced physiological state. Butyrate is a short-chain fatty acid produced by gut bacteria through the fermentation of dietary fiber. Butyrate supports gut health by maintaining the integrity of the gut barrier, reducing inflammation, and promoting the growth of beneficial bacteria. Supplements that increase butyrate production, such as resistant starch, can be beneficial in managing dysbiosis.

Digestive Enzymes help break down food more efficiently, aiding in nutrient absorption and reducing the burden on the gut. Products like Integrative Panplex 2-Phase gastric and intestinal digestive support provide enzymes like pepsin, L-glutamic acid HCL, and pancreatin (amylase, protease, and lipase) that improve digestion and reduce the likelihood of undigested food reaching the lower gut, where it could promote the growth of harmful bacteria.

Inulin is a type of prebiotic fiber found in various plants. Inulin promotes the growth of beneficial bacteria like *Bifidobacteria* in the gut, helping restore balance and counteract dysbiosis.

L-Glutamine is an amino acid that plays a key role in gut health, supporting the integrity of the gut lining and helping reduce gut permeability (leaky gut), which can be a consequence of dysbiosis. A healthy gut lining helps maintain a balanced microbiome.

Zinc Carnosine is a combination of zinc and carnosine that supports gut health by maintaining the integrity of the gut lining and the immune system, making it effective in managing dysbiosis and preventing harmful bacteria from proliferating.

Saccharomyces Boulardii is a beneficial yeast that acts like a probiotic. S. boulardii can help restore balance in the gut microbiome by inhibiting harmful bacteria and supporting the growth of beneficial bacteria. It is particularly useful in managing diarrhea and gastrointestinal disorders associated with dysbiosis.

Garlic Extract (Allicin) is a natural antimicrobial compound found in garlic. Garlic extract, particularly allicin, has antimicrobial properties that can help reduce harmful bacteria in the gut, making it a useful supplement for managing dysbiosis.

Other Key "Master Supplements"

I take a few other supplements daily that I think of as "master supplements" not listed under the twelve hallmarks above. They include:

Glutathione is a naturally occurring tri-peptide amino acid made from cysteine, glycine, and glutamine produced in every cell of our body. Known as the master antioxidant, glutathione is vital for overall health. It is the body's first line of defense against free radical stressors our bodies encounter daily, like pollution, UV rays, chemicals, and more. However, aging, oxidative stress, and toxins lessen our glutathione levels. Glutathione is particularly effective in addressing many of the hallmarks of aging including mitochondrial dysfunction, genomic instability, loss of proteostasis, cellular senescence, chronic inflammation, altered intercellular communication, stem cell exhaustion, and epigenetic alterations. There are three options to maximize your absorption of glutathione: First, a liposomal formulation that is taken orally; second, glutathione delivered via an intravenous drip, typically as part of health-related IV treatment; and third, a spray-on glutathione formulation, which maximizes absorption. (I use this product from Aurowellness.com. **Note:** I have no relationship with Aurowellness.) Glutathione detoxifies cells, combats oxidative stress, and supports liver function, helping maintain robust cellular health and enhancing immune defenses.

Stamets-7: A unique blend of seven medicinal mushrooms, including Reishi, Lion's Mane, and Cordyceps, that work synergistically to support immune function, cognitive health, and energy levels. This supplement also promotes resilience against stress and inflammation, contributing to overall vitality and well-being. These mushrooms work synergistically to address many of the hallmarks of aging, including mitochondrial function, chronic inflammation, altered proteostasis, decreased intercellular communication, declining stem cell function, altered nutrient sensing, and genomic instability.

Polyphenol Booster: A supplement rich in polyphenols, polyphenol boosters are naturally occurring compounds with strong antioxidant and anti-inflammatory properties. They help protect against oxidative stress, support cardiovascular health, and promote metabolic balance, contributing to overall wellness. This supplement addresses multiple hallmarks of aging, including enhancing mitochondrial function, reducing inflammation, maintaining proteostasis, preserving stem cell health, and protecting against genomic instability.

Creatine: A well-known supplement that supports muscle strength, power, and endurance by enhancing ATP production in muscles. It improves exercise performance, supports muscle recovery, and helps maintain overall physical function. Creatine addresses many of the hallmarks of aging, addressing, first and foremost, mitochondrial dysfunction through its support of ATP production. Creatine has also been shown to protect cells from oxidative stress, one of the triggers for cellular senescence. It also has anti-inflammatory properties and can reduce the production of pro-inflammatory cytokines. It also supports immune function, helping modulate the inflammatory response. Finally, by reducing oxidative stress, creatine can help protect DNA from damage, which is a major contributor to genomic instability. I consume 5 mg of creatine daily in one of two forms, either as a powder mixed into my protein shakes or creatine gummies (a product called Create) when traveling.

Taurine is an amino acid with a wide range of pro-longevity benefits. To begin, taurine reduces oxidative damage to DNA, proteins, and lipids, helping maintain cellular integrity and function. It also maintains mitochondrial health by stabilizing mitochondrial membranes and supports the production of ATP. Taurine helps regulate blood sugar levels by improving insulin sensitivity and supporting glucose metabolism. Finally, taurine supports cardiovascular health by regulating blood pressure, improving lipid profiles, and protecting against heart disease—reducing the risk of atherosclerosis by preventing the accumulation of cholesterol in the arteries.

Supplements Supporting Brain Health

Without question, at the top of my health goals is maintaining excellent brain health (memory, cognitive function, and alertness) and ultimately preventing or at least delaying any dementia until breakthroughs can prevent it altogether. The following are the additional brain-health-related supplements I take every day:

Prodrome Neuro supports brain health primarily by promoting the synthesis of omega-3 plasmalogens, an essential phospholipid for the structural integrity of cell membranes. By protecting against oxidative stress and neuroinflammation, supporting neurotransmitter function, and aiding in the repair and maintenance of brain cells, Prodrome Neuro helps preserve cognitive function and protect against neurodegenerative diseases.

Prodrome Glia supports brain health by restoring omega-9 plasmalogen levels and preventing demyelination and neurodegeneration by maintaining the integrity of the myelin sheath. Prodrome Glia plays a crucial role in promoting brain health and protecting against age-related cognitive decline by reducing lipid peroxidation.

Liposomal Phospholipid Complex supports brain health by enhancing the integrity and function of neuronal membranes, improving neurotransmitter function, protecting against oxidative stress, maintaining myelin sheath integrity, and supporting the blood-brain barrier. By delivering these benefits in a bioavailable form, the complex helps preserve cognitive function, protect against neurodegenerative diseases, and promote overall brain health as we age.

BAET or Beta-AET (androst-5-ene-3β,7β,17β-triol) is a metabolite of DHEA (dehydroepiandrosterone) and has notable anti-inflammatory and immune-stimulating properties. Beta-AET can help mitigate neuroinflammation by reducing these pro-inflammatory cytokines, potentially protecting neurons from damage and reducing the risk of cognitive decline and neurodegeneration. Beta-AET is also known to enhance immune function, which is crucial for protecting the brain from infections and other environmental insults that can lead to cognitive impairment.

Supplements Supporting Cardiovascular Health

As I mentioned earlier, the highest mortality risk comes from cardiovascular disease. Following are the heart-health-related supplements I take every day:

Arterosil is a supplement that has been shown to regenerate the endothelial glycocalyx, which is the protective lining inside our blood vessels. It thus supports endothelial function, promoting proper blood flow and reducing the risk of cardiovascular disease, which may contribute to overall longevity and support heart health.

Other supplements in this area, some mentioned above, include **berberine**, a natural compound found in several plants that has been shown to support healthy blood glucose and lipid levels, reducing inflammation and improving endothelial function. **Niacin (Vitamin B3)** is well-known for its ability to improve lipid profiles by increasing HDL (good cholesterol) and reducing LDL (bad cholesterol) and triglycerides. **Omega-3 Fatty Acids (EPA and DHA)** are essential for reducing inflammation, lowering triglycerides, and supporting overall cardiovascular health. **Coenzyme Q10 (CoQ10)** is a powerful antioxidant that supports energy production in the heart muscle, reduces oxidative stress, and improves endothelial function. **Taurine** regulates calcium levels in heart cells, reducing blood pressure and improving cholesterol profiles. **Vitamin K2 (MK-7)** plays a crucial role in directing calcium into the bones and preventing its deposition in the arteries, reducing the risk of arterial calcification. **Garlic Extract (Allicin)** has been shown to reduce blood pressure, improve cholesterol levels, and reduce arterial stiffness.

New supplements in the heart-healthy category include:

L-Arginine is an amino acid that serves as a precursor to nitric oxide (NO), a molecule that helps relax blood vessels, improve blood flow, and reduce blood pressure. Improved nitric oxide production supports overall cardiovascular health.

Pomegranate extract is rich in polyphenols, which have powerful antioxidant and anti-inflammatory properties. It helps reduce oxidative stress, lower blood pressure, and improve cholesterol levels, contributing to better cardiovascular health. The right microbes can convert pomegranate to urolithin A in the gut.

NAD+ BOOSTERS: NMN, NR, and Nuchido TIME+

One particular group of supplements—NAD+ (Nicotinamide Adenine Dinucleotide) boosters—deserve their own deep dive, given their growing importance in longevity. Before I share with you what I'm taking and why, let's dive into some important background information on NAD+ and why boosting this particular molecule in your body is important.

Let's begin with sirtuins. Sirtuins are a set of seven regulatory proteins that have two different and competing functions in your cells. First, they govern our epigenome, turning the right genes on at the right time within the right cell types. They boost mitochondrial activity, reduce inflammation, and protect our telomeres. Second, they have another critical function in directing DNA repair. All in all, these proteins and the genes that code for them are damn important.

As we age, the need for DNA repair increases because of accumulated damage. Makes sense, right? At age 20 we've only had a little exposure to environmental toxins and radiation. However, by age 60, we have had three times the exposure, and because DNA damage accumulates, the need for repair is constantly increasing. Therefore, our sirtuins become overtaxed, frantically responding to one fire alarm after the next. As they focus on DNA repair, they get distracted from their second very critical job of regulating our epigenome, deciding which genes should be turned on and which should be off.

As a result of this over-taxation of the sirtuin system, our ability to repair the damage while regulating the epigenome becomes increasingly challenging as we age and accumulate more DNA damage. From entire organ systems to individual cells, our bodies become dysregulated. Epigenetic noise accumulates. Genes that have no business being on are allowed to turn on, and genes that should be on are allowed to turn off. It's epigenetic mayhem!

In a nutshell, that's the dynamic of aging on a molecular level: *the tension between gene regulation and gene repair* and how we pay the price when our sirtuins become overwhelmed.

If that were the end of the story, it would be challenging enough, but it gets worse. The fuel that enables sirtuins to do their critical work diminishes as we age.

The Critical Role of NAD+

Thanks in large part to Dr. David Sinclair's work at Harvard Medical School, we now know our sirtuins can't do much of *anything*—including fixing our DNA—without a steady supply of NAD+, a molecule that is critical to power the entire sirtuin system. So, it's sobering to learn that we lose about half of our NAD+ by our fifties—right around the time we need it more than ever to function at peak efficiency. Not only do our sirtuins have more work to do as we age, but they also don't have enough NAD+ fuel to do their job. Does that sound like a challenge for anyone over 50? Luckily, there's something we can do about it.

NMN and Muscular Mice

In his research at Harvard, Sinclair conducted a remarkable experiment. His team gave NMN (nicotinamide mononucleotide), a molecule that is converted into NAD+ inside our cells, to twenty-month-old mice (equivalent to humans in their sixties). The results were transformative, reviving their mitochondria and increasing their blood flow and the size and strength of their muscles. Within two months, the revitalized animals ran 60 percent farther than an untreated control group. They'd become as vigorous as mice half their age. By every measurement that mattered, they were young again. This is why Sinclair takes a gram of NMN every morning as a supplement.

How I Supplement My NAD+

When I think about boosting my NAD+ levels, there are three approaches I've evaluated and used. I tend to switch between all three of these and have lately endeavored to obtain accurate measures of my intracellular NAD+ levels (which are fairly difficult to measure). It's important that you research this yourself and determine which approach is best for you.

First: Taking Oral NMN—Finding a Supplier

In years past, nicotinamide mononucleotide was widely available online, and the challenge was choosing between hundreds of sources to select a brand of NMN from a supplier delivering a pure and stable product. However, recently, this has all changed.

In November 2022, the FDA issued a statement *declaring NMN cannot be marketed as a dietary supplement* in the United States. The FDA's decision was based on the fact that NMN had recently come under investigation as a drug. So, as of now, NMN is considered a drug candidate, which means it cannot be sold or marketed as a dietary supplement in the US, and companies have been required to stop sales and remove their products from the market.

In that same year, 2022, I was introduced to a company called Elevant by Eric Verdin, PhD, CEO of the Buck Institute. Based in Europe, Elevant manufactures Optima, which delivers NMN in pleasantly tasting 125 mg chewable tablets. I typically take between 750 to 1,000 mg per day. (**Note:** I have no financial relationship with Elevant.)

According to the science team at Elevant, one of the key properties that positions NMN as an effective NAD+ booster is its potential speed of absorption into the body. In 2019, Elevant chief scientist Dr. Alessia Grozio was part of a team that published research that highlighted a key driver of NMN's efficiency—a transporter that carries NMN directly into cells within minutes after ingestion.[93] It was demonstrated that the SLC12A8 gene encodes a protein that is a specific NMN transporter in mammals. This protein uses a sodium ion to transport NMN (but not NR) across cell membranes and facilitates direct uptake of NMN into the gut and other organs. Biosynthesis significantly increases NAD+ concentration in the cells over the course of sixty minutes following ingestion. This indicates that NMN represents a direct entry point into the NAD+ synthesis pathway.

Special Forces and Their NMN Experiments

One company, Metrobiotech, is currently conducting a human clinical trial of a crystalized version of NMN named MIB-626 (which differs in the way it is manufactured but not in its chemical composition or function). These trials are looking at the impact of MIB-626 in a wide variety of indications, ranging from increased muscle endurance and neurogeneration to treating COVID-related kidney failure and even heart failure. If Metrobiotech is successful in its research and proposal to the FDA, NMN will be characterized as a drug and sold as a prescription. Perhaps most interestingly, in July 2021, it was leaked that the US Special Operations Command (SOCOM) had "completed preclinical safety and dosing studies in anticipation of follow-on performance testing" using Metrobiotech's MIB-626 molecule.[94] "If the preclinical studies and clinical trials bear out, the resulting benefits include improved human performance, such as increased endurance and faster recovery from injury," said Navy Cdr. Timothy A. Hawkins, a spokesperson for SOCOM. If all goes well with clinical trials, it is hoped that MIB-626 will gain regulatory approval as a new drug available to all of us by the end of 2025 or 2026.

Second: Boosting NAD+ with NR (Nicotinamide Riboside)

NR is a naturally occurring form of Vitamin B3 that can be converted into NAD+. Both NR and NMN are intermediates in the NAD+ biosynthetic pathway. That means they're both steps in the body's process to produce NAD+. They're closely related in that they both aim to increase NAD+ levels, though they do so through slightly different pathways.

Third: Boosting NAD+ with Nuchido TIME+

Nuchido TIME+ includes a number of our superstar supplements, including Sophora japonica tree extract, alpha-lipoic acid (ALA), apigenin (which inhibits a protein called CD38) derived from parsley, epigallocatechin-3-gallate (EGCG) from green tea extract, vitamin C, zinc, and black pepper extract. The product works to increase NAD+ levels by providing precursors for NAD+ synthesis, inhibiting NAD+ degradation, enhancing the NAD+ salvage pathway, supporting mitochondrial health, and reducing oxidative stress. These combined actions help restore and maintain NAD+ levels. I take three capsules in the morning and three in the evening when I use the product. (**Note:** I have no relationship with Nuchido TIME+.)

YOUR LONGEVITY PHARMACY

My Skincare Routine

Longevity scientist Robert Hariri, MD/PhD, says: "For longevity to succeed, it needs to support esthetic, mobility, and cognition. You need to look good, move well, and think clearly." There's no doubt that looking at yourself in the mirror and feeling confident about the appearance of the person looking back at you is critical to a longevity mindset. And it's for that one reason that skin care is an important part of everyone's healthspan strategy.

It feels odd that I'm writing this. I never thought I'd have a skincare routine, let alone write about one. This all changed when I found a product that helped me reverse the age of my skin. If you know me, you know that I'll try just about anything (within reason—sometimes). I figured it was worth a shot, and since using it, I've received more compliments on my skin than ever (seriously).

OneSkin's products can reverse skin's biological age by decreasing the level of senescent cells.

The company is called OneSkin, and I use their product, OS-O1 Face. This isn't an ad, but I am an advisor to the company. The science of OneSkin is fascinating and worth discussing for a moment.

OneSkin was cofounded in 2016 by a pioneering team of four female PhDs from Brazil. They include CEO Carolina Reis Oliveira (an expert in stem cell biology) and Chief Science Officer Alessandra Zonari (an expert in skin regeneration), who dazzled the audience at my Longevity Platinum Trip in 2020.

Their mission is to revitalize your skin from the inside and extend your "skin-span"—the time your skin remains healthy and youthful. How? "We're targeting what we believe is the root cause of skin aging," says Dr. Oliveira. As she explains, the chief culprits are senescent cells, which are damaged cells that build up in the body, contributing to aging and age-related diseases. As senescent cells accumulate in our skin, they create wrinkles and sagging, cause inflammation, and make us more susceptible to skin cancer. So OneSkin sets out to kill them.

The company developed a powerful screening platform that made it possible to evaluate about one thousand small peptides (short amino acid sequences) to see if any of them could eliminate senescent cells. They hit the jackpot, discovering one highly effective 20 amino acid peptide that they christened OS-01. OneSkin's experiments have shown that their peptide can significantly decrease the level of senescent cells, reducing the age of the skin by several years at a molecular level.[95] In short, it's about long-term restoration and rejuvenation.

I use the product twice per day, morning and night.

And yes, I also use their SPF 30 product on my face for sun protection, especially on days when I'm out and about in the sun.

Are there other skin lotions and potions that make similar claims? I'm sure there are, but I'm super happy with OneSkin and, frankly, haven't looked elsewhere.

Therapeutic Treatments *(I'm exploring)*

Therapeutic treatments or protocols are interventions performed under a physician's care and sometimes under an investigational new drug (IND) protocol and/or an Institutional Review Board (IRB), which serve different but complementary roles in therapeutic research.

An IND is an application to the FDA to test a new drug, biologic, or treatment not yet FDA-approved and is obtained before the start of clinical trials. An IRB is a committee that reviews and monitors research involving human subjects, focused on protecting human research participants' rights, safety, and welfare, including ethical aspects of study protocols, informed consent, and risk management.

In the remainder of this chapter, I want to share therapeutic treatments I have either participated in or am considering through Fountain Life. This list will grow as new protocols become available later this decade.

Senolytics Protocol

As mentioned in the discussion of OneSkin and OS-01, a senolytic peptide for the skin, senolytics are a class of drugs that selectively target and eliminate senescent cells, which are cells that have stopped dividing and have entered a state of permanent growth arrest. Senescent cells accumulate with age and can contribute to aging and age-related diseases by secreting pro-inflammatory cytokines, growth factors, and proteases, collectively known as the senescence-associated secretory phenotype.

There has been growing interest in senolytics as potential therapies for improving healthspan and lifespan. While the field is still emerging, several senolytic agents have gained attention for their potential longevity benefits:

1. **Dasatinib and Quercetin**: As mentioned earlier, this combination is perhaps the most studied senolytic regimen. Dasatinib, originally a cancer drug used to treat certain types of leukemia, and quercetin, a natural flavonoid found in many fruits and vegetables, have been shown in preclinical studies to selectively eliminate senescent cells in various tissues when given as an intermittent dosing regimen, typically two to three days in a row once or twice a month.[96] Dasatinib requires a prescription, and this protocol should be conducted with medical oversight.
2. **Fisetin**: A natural flavonoid found in fruits like strawberries, fisetin has demonstrated senolytic properties in both cell culture and animal models.[97]
3. **Spermidine**: A natural polyamine, spermidine has been shown to extend lifespan in various model organisms, possibly through autophagy activation and potential senostatic effects.[98]

It's important to note that while many of these compounds show promise in preclinical studies, their efficacy and safety in humans still need to be fully established, especially for long-term use. Clinical trials are underway for several of these agents to assess their potential as senolytic therapies in humans.

It is important to remember that you don't want to eliminate all senescent cells as they also have a beneficial role to play in your body. For example, they are important in wound healing and tumor suppression. Ideally, you should measure your level of senescence to determine if you may benefit from senolytic therapy. I have recently used an IRB-approved research test called SapereX from Sapere Bio to measure my cellular senescence. Having found my cellular senescence is overall low, I will not be pursuing any therapeutics in this area, but I will continue to test my cellular senescence in the future.

Using NK Cells as a Senolytic

Another approach, pioneered by Dr. Robert Hariri at Celularity (of which I am a co-founder), is eliminating senescent cells by supplementing a patient's natural killer (or NK) cells. NK cells are a class of lymphocytes, part of your innate immune system, that are an early defense against infections and the prevention of cancer. They also help identify and eliminate senescent cells. Celularity is currently developing a number of therapeutics utilizing placental-derived NK cells to augment a patient's existing population. The company is working to achieve this by extracting your NK cells, expanding them in number, and returning them to you. This is currently being developed under a research protocol. Once this treatment is available under an investigational new drug protocol, I will consider using it—not to reduce my senescent cell count, which is already low, but for the benefits of supplementing my NK cell count to detect and fight nascent cancers and prevent infections.

Therapeutic Plasma Exchange (TPE)

Have you heard of parabiosis? If you connect an old mouse's circulatory system with a young mouse, the old mouse gets younger.

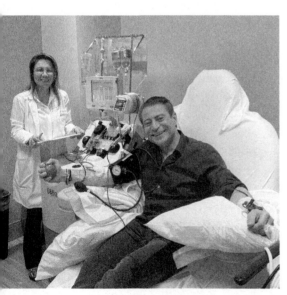

Peter receiving Therapeutic Plasma Exchange (TPE) at Orlando Fountain Life Center.

The concept was popularized in the HBO series *Silicon Valley*, in which a tech billionaire uses a young "blood boy" as a source of young plasma transfusions for his longevity treatments.

Aside from the moral issues, the rejuvenation results were shocking in the case of the mice. A variety of the old mouse's tissues, organs, and fur had the characteristics of a far younger, healthier mouse.

Follow-up studies confirmed this finding and showed that the opposite was also true. Transfuse younger animals with blood from older ones, and the clock spins forward, accelerating decrepitude and amplifying aging.

In a new twist called therapeutic plasma exchange, achieving these same benefits in humans might be possible without needing your very own "blood boy." Recent clinical human trials using TPE have already reported positive results for the treatment of various age-related diseases, such as Alzheimer's—slowing disease progression by over 65 percent.[99]

What Exactly Is TPE, and How Does It Work?

Therapeutic (or total) plasma exchange is a medical procedure that removes the liquid part of the blood (plasma), which is the portion that contains proteins, antibodies, toxins, inflammatory agents, and other substances, and replaces the old plasma with fresh saline and albumin. Historically, TPE has been successfully used to treat various diseases, including autoimmune diseases, neurological disorders, hematological disorders, and poisoning.

In the case of Alzheimer's, the AMBAR clinical trial has shown that patients treated with TPE experienced 66 percent slower disease progression than their control group peers. Additionally, a study published in 2021 found that 78.7 percent of multiple sclerosis patients showed clinical improvement after TPE.[100] Now, the question is, *can TPE help with longevity and increase healthspan?*

Can TPE Improve Longevity and Healthspan?

In recent studies, it's been suggested that as we age, our blood accumulates harmful substances (like cytokines and toxic autoreactive antibodies), which could be removed via the TPE process. One study found that TPE increased the lifespan of mice by 30 percent.

Dr. Dobri Kiprov, an internationally renowned pioneer and expert in therapeutic apheresis, immunotherapy, long COVID, and age-related disorders, was the first to publish about the effect of TPE on aging.

In 2022, Dr. Kiprov teamed up with his scientific partner, Dr. Irina Conboy, to publish a landmark study in the journal *GeroScience*, showing that TPE can indeed have positive rejuvenation effects.[101] The study revealed that human biological aging and many of the debilitating conditions that come with it are driven by excesses of molecular bad actors that accumulate in the blood as people age. These circulating blood proteins include cytokines, toxic autoreactive antibodies, and biomarkers for specific diseases.

The researchers studied the effects of multiple rounds of TPE on people of varying ages.

Their findings showed indicators of rejuvenation in the samples, including reduced inflammaging, diminished protein markers of neurodegeneration, and improved immunity.

In a recent conversation with Dr. Kiprov, he offered this advice and recommendation:

> **"We recommend six treatments because that has been our experience so far in a controlled clinical environment. After the six treatments, all the parameters improved dramatically. The improvements include a reduction in (epigenetic) aging clocks between two to five years and decreased muscle loss."**

Dr. Conboy suggested that there is a preventive capacity against age-associated neurological disease as well. She says, "By understanding how rapid and drastic dilution of age-elevated proteins in blood serum works, we can eventually make all tissues and organs younger."

Why I'm Using TPE, and Is the Cost Worth It?

I refer to my TPE treatments as my "longevity oil change." Just last week, at the time of this writing, I completed five of six recommended treatments utilizing the TPE machine at the Orlando Fountain Life Center.

TPE treatments work by removing around 2 to 3 liters of old plasma and, following Dr. Kiprov's protocol, replacing the old plasma, which contains "pro-inflammatory and aging factors," with a mixture of 5 percent human albumin (collected from blood donations at blood banks and plasma donation centers) and roughly 95 percent saline.

My experience with TPE has been relatively smooth and easy. The entire process takes about three hours from start to finish. During the process, I'm usually on Zoom or taking phone calls. One of my friends put on his Apple Vision Pro headset and watched a movie.

After breakfast and hitting the restroom, I arrive early, about 9:00 a.m. The nurse puts one needle in my left arm that extracts blood, flowing it into the TPE machine where the red and white blood cells and platelets are separated from the plasma and eventually returned to my body. Another needle is placed in my right arm, and the fresh saline, albumin, and retained red and white cells are returned to my circulation. On my fourth TPE treatment, when I asked the nurse to run the process faster so I could get to a scheduled podcast, I felt woozy and hypoglycemic. I quickly recovered when she slowed the rate and gave me juice to raise my blood sugar. In general, the afternoon after the treatment leaves you a bit tired—similar to what it feels like after donating blood—but typically, I'm feeling great 24 hours later, clear-headed and energetic.

After I finish my sixth TPE treatment this fall, I'll repeat my TruAge tests and see if the treatments have impacted my aging clocks. TPE treatments are not cheap; they range in price from $8,000 to $12,000 per treatment (though I expect the price to come down over time), and as of writing this, I can say that the subjective results of these treatments thus far have been unclear. However, for those with diseases such as autoimmune neurological disorders, thrombotic microangiopathies, hyperviscosity syndromes, autoimmune hematological disorders, severe systemic autoimmune diseases, and possibly even Alzheimer's, TPE can be a lifesaving procedure.

The Power of Stem Cells

One of the hallmarks of aging discussed earlier is stem cell exhaustion, which occurs as we age when the population of stem cells in our body diminishes by orders of magnitude. But what if you could supplement and replenish your stem cell supply later in life? That's the next therapeutic up for discussion.

By some accounts, we are at the cusp of a stem cell revolution. Understanding and harnessing these unique cells may unlock breakthroughs in longevity and therapeutic solutions for *all kinds* of chronic diseases and regenerative opportunities. Over the last decade, the number of publications per year on stem cell-related research has increased forty-fold. The global stem cell therapy market is expected to *nearly triple* from $11.2 billion in 2022 to over $31 billion by 2030.

But what exactly are stem cells? And what's possible with safe and effective stem cell treatments?

Stem cells are undifferentiated cells resident in every tissue and organ in your body and can differentiate into specialized heart, neuron, liver, lung, or skin cells. Beyond their ability to become any cell type, they can divide to produce more stem cells. In a child or young adult, stem cells are in large supply, acting as a built-in repair system, and are often summoned to the site of damage or inflammation to repair and restore normal

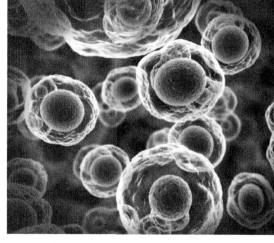

Stem Cell Exhaustion: Humans lose 100x – 1000x of our stem cells as we age.

function. However, as we age, our supply of stem cells begins to diminish, as much as 100 to 10,000-fold, in different tissues and organs.

An analogy shared with me by Robert Hariri, MD, PhD, might help illustrate their role in aging. "Imagine your stem cells as a team of service technicians in your newly constructed mansion. When the mansion is new and the technicians are young, they can fix everything perfectly. As the technicians age and reduce in number, your mansion goes into disrepair and eventually crumbles."

But what if you could restore and rejuvenate your stem cell population?

Let's begin by understanding the two major sources from which we might access stem cells for this purpose. First are autologous stem cells, which are derived from you, typically from your fat or bone marrow, extracted and given back to you, or stored and delivered back to you at a future date. In other words, this involves you receiving your own stem cells. The second broad source of stem cells is allogeneic stem cells, which come from someone else, a donor, often from either the umbilical cord or placental stem cells that are usually discarded during birth.

It's important to know that today, the use of allogeneic stem cells is not approved by the FDA and is, therefore, not legal in the United States. For this reason, many people travel to permissive jurisdictions outside the US, such as Mexico, Antigua, Panama, or Costa Rica. I'm often asked by friends where they should go for treatments. After evaluating many venues for their medical and scientific safety, I found the Regenerative Medicine Institute (RMI) in Costa Rica to have strong medical support and GMP facilities. RMI offers a wide range of services, including access to allogeneic umbilical cord stem cells for joint repair and general IV infusion. **Note:** I don't have an official relationship with RMI; I am simply a fan of their work.

While not on the immediate horizon, there is another type of stem cell worth discussing that will be useful and accessible during the second half of this decade, something called an induced pluripotent stem cell or iPSC.

iPSCs represent a groundbreaking advancement in regenerative medicine and stem cell research, originating from the pioneering work of Nobel Laureate Shinya Yamanaka, PhD, and his team in 2006. Yamanaka discovered that by introducing a specific set of four transcription factors—OCT4, SOX2, KLF4, and c-Myc—into adult somatic cells, such as skin cells, these cells could be reprogrammed into a pluripotent state. This reprogramming process essentially resets the adult cells, transforming them into cells that resemble embryonic stem cells, which have the ability to differentiate into any cell type in the body. iPSCs are manufactured in the laboratory by carefully delivering these transcription factors into the target cells using viral vectors or other gene delivery methods. Once reprogrammed, iPSCs can be induced to differentiate into any type of cell—hepatocyte, neuron, muscle, skin, etc.—offering a versatile platform for studying disease, developing personalized therapies, and potentially regenerating damaged tissues, all while avoiding the ethical concerns associated with the use of embryonic stem cells.

Dean Kamen and His ARMI: Using Stem Cells to Regrow Your Organs

What if you had a supply of "spare organs"—hearts, livers, lungs, and kidneys—available to you as you age? This is just one of the incredible technological developments that has the potential to extend our healthspan that is being created at the Advanced Regenerative Manufacturing Institute (ARMI).

In the United States alone, over 100,000 people are waiting for an organ transplant, and every 10 minutes, someone is added to the list. *On an average day, 17 people die waiting for an organ.* Poor organ availability represents one of the biggest failures of our healthcare system.

Imagine this: By 2030, a factory in New Hampshire (where ARMI is located) collects skin cells from you, converts them into iPSCs, grows them into billions of cells in a bioreactor, and finally induces them to differentiate to manufacture cells for a heart, liver, lung, or kidney for you. They could create whatever organ you desire. This is the vision of Dean Kamen and his team at ARMI.

Dean Kamen at ARMI building spare organs.

ARMI's mission is to facilitate the manufacturing of human tissue for organs and to automate and democratize the process—at scale. Kamen believes that ARMI will be the "birthplace of the next great industrial growth spurt in the world": one for manufacturing organs. ARMI recently secured an additional $100 million from the US Department of Defense, bringing its current level of capital to well over $200 million.

Every other year, in 2021 and 2023 (and again in 2025), I bring a group of Abundance Platinum members to visit Dean's ARMI factory and tour his famed New Hampshire home (a cross between Willy Wonka's Chocolate Factory and Lockheed's Skunkworks). This past year, ARMI's progress was breathtaking; they expanded their manufacturing footprint by one hundredfold and began developing and producing tissues to address a range of medical needs.

What medical needs are they building organs to address? First on the list is the creation of pancreatic beta cells to treat diabetes, a medical condition that Dean Kamen has been focused on solving for decades. In collaboration with the Juvenile Diabetes Research Foundation (JDRF), ARMI is building bioreactors to mass-produce islet cells for diabetes research so ARMI can start building massive quantities of islet cells to give to every researcher that JDRF funds across the US. In 2024, ARMI will be shipping some of these islet cells around the country.

Next, in partnership with Doris Taylor, PhD, who moved to ARMI from the Texas Heart Institute, the team at ARMI has set out to build miniature, pediatric-scale hearts. After all, should a child need a heart transplant, where would you get a donor heart?

However, this is just the beginning. Dean and his team are also developing iPSC-derived skin, bone, and ligament segments and lung, kidney, and retinal cells to address macular degeneration.

Best of all, no immunosuppression drugs are required for the recipients of these manufactured organs. Why? Because the recipient's own skin cells are used to create the iPSCs, which are grown and differentiated into the desired tissues. As such, the iPSC-derived organs are ultimately built from the recipient's own DNA.

What I Am Doing About Stem Cells and BioBackup

The entire field of stem cell treatments can be somewhat confusing, and I recommend using caution. The good news is that the field is moving rapidly, and I'm hopeful that regulatory clarity and strong validating human trials will materialize within the next few years.

So what am I doing? The first thing I've done is back up my biology and bank my stem cells and immune cells (even though they are 63 years old, they will never be younger than they are today). I'm doing this through a service called "BioBackup," which is delivered by one of my companies, Celularity, under the guidance of Celularity's CEO, Bob Hariri.

Just like backing up your computer data, backing up your biology may have long-term value. The cells are acquired through a simple process of collecting peripheral blood (the same as going to donate blood at the Red Cross). Your circulating blood contains many immune cells and small quantities of stem cells, which can be isolated and cryopreserved for future use. Cryopreserved cells are viable for decades without degrading their biological potential and can be revived and used at regular intervals throughout your lifetime. Celularity already does this for newborns, collecting umbilical cord blood and placental cells and storing them for decades through their LifeBankUSA division. They've banked over 100,000 newborn placentas, including the cells of my twin boys when they were born thirteen years ago.

The second thing I hope to do is to avail myself of Celularity's placental stem cells once they become available through an FDA-approved IND (investigational new drug) protocol.

The final topic related to stem cells is the subject of exosomes, a technology I've used with great success.

Exosomes and How I've Used Them

Exosomes are small, membrane-bound vesicles secreted by almost all human cells and play a crucial role in cell-to-cell communication. They are typically 30 to 150 nanometers in size and contain a rich cargo of proteins, lipids, RNA, and other bioactive molecules. Exosomes are formed within the endosomal compartment of cells and are released into the extracellular environment when multivesicular bodies fuse with the plasma membrane. Once released, exosomes can travel through bodily fluids, such as blood, saliva, and cerebrospinal fluid, to reach and interact with distant cells. By transferring their molecular cargo, exosomes can influence the behavior of recipient cells, modulating processes like immune response, inflammation, and tissue repair.

Exosomes, specifically those from stem cells, have gained significant attention in both research and clinical settings due to their potential as biomarkers and therapeutic

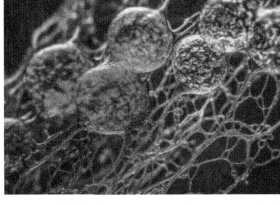

Exosomes are a promising experimental avenue for human rejuvenation.

agents. Exosome-based therapies are being developed to promote tissue regeneration to enhance healing and repair processes. Exosomes have been used to address chronic pain, osteoarthritis, and musculoskeletal injuries. Note: the FDA does not currently approve exosomes, and their use should be subject to an IND research protocol.

I've used exosomes in several ways. First, and most effectively, I utilized them post-surgery to quicken tissue repair. Here's a brief story that gives me some confidence in their efficacy. About thirteen years ago, I underwent a right shoulder rotator cuff repair without the use of exosomes (a technology I wasn't quite aware of yet). Nearly a decade later, I had the same surgery, using the same surgeon on my left shoulder. This time, however, aware of the regenerative potential of exosomes, I had them injected into my shoulder joint at the site of repair on two occasions, one week and two weeks post-surgery. While my reported experience is obviously subjective and not fully quantified, it was clear to me that recovery of the shoulder that utilized exosomes was substantially faster and less painful. And that was the shoulder of someone eleven years older.

I've also used exosomes intravenously and injected them into my scalp to stimulate (follicular) hair growth under IND studies.

<u>Disclaimer:</u> *There are risks associated with exosome and stem cell-based therapies, as these treatments may stimulate the proliferation of pre-existing cancers. It is crucial to undergo comprehensive diagnostic testing before considering these therapies to establish a baseline awareness of any pre-cancerous conditions or existing tumors. Such conditions must be thoroughly addressed and treated by qualified medical practitioners before participating in any therapies or experimental protocols discussed in this book.*

Insights from Hans Keirstead, PhD: Stem Cell Secretome Therapy for Longevity

At my 2024 Longevity Platinum Trip, Hans Keirstead, PhD, chairman of Immunis, introduced his cutting-edge therapeutic, IMMUNA, for restoring immune function and regenerating muscle's overall vitality. IMMUNA utilizes a stem cell-derived secretome—a well-characterized mixture of proteins, exosomes, and growth factors released by stem cells—to reverse the age-related decline.

Hans Keirstead, PhD

Stem Cell Secretome: Precision and Potential

Keirstead's lab has achieved high purity in the differentiation of stem cells, meaning they can *consistently* produce the youngest, most potent cells for their IMMUNA product. These stem cells secrete a wide array of bioactive molecules; he's identified 440 factors that stimulate immune cells to grow, communicate more efficiently, and navigate through tissues more effectively. "We're able to multiply these cells in the lab by billions," Keirstead explained, "removing batch-to-batch variability, so every patient gets the same high-quality treatment."

Human Phase: 1/21 Clinical Trial Results

Building on very successful preclinical animal studies, Immunis conducted a US FDA-approved Phase 1/2a clinical trial (NCT05211986) in elderly individuals aged 50 to 75 who were suffering from muscle atrophy and were immobilized secondary to knee osteoarthritis (OA). The trial, which tested the safety and efficacy of IMMUNA, *revealed no adverse events* attributed to the treatment. Of particular note, participants in this trial experienced a 6 percent increase in muscle size without exercise (remember, these OA patients were generally immobilized), and their muscle strength also improved. Most notably, their walking ability, as measured by the six-minute walk test, increased by 10 percent in just four weeks and *increased by 150–200 percent over three months*! This functional improvement in gait speed and distance surpassed the outcomes of age-matched controls who exercised twice a week for six months, showcasing the potential of IMMUNA to outperform exercise in restoring mobility and strength.

The therapy also resulted in significant reductions in pain, with patients reporting a 50–70 percent decrease. Inflammatory markers such as TNF-alpha, GM-CSF, IL-6, IL-8, and IL-10 were significantly reduced, indicating a dramatic decrease in chronic inflammation. Additionally, the treatment nearly eliminated markers of cartilage degeneration, which are common in aging populations. These results highlight IMMUNA's potential to restore muscle and metabolic function and address inflammation and joint degeneration.

In September 2024, the FDA granted clearance for Immunis's follow-up Phase 2 clinical trial, which will assess muscle gain and fat loss in a broader patient population suffering from sarcopenic obesity. This trial will utilize an adaptive clinical trial design to ensure the therapy continues to be refined and optimized as it progresses through the clinical pipeline. The overarching objective remains to restore the immune system's functionality to that of a healthy 27-year-old, potentially reversing the age-related decline in muscle mass, metabolism, and overall vitality.

Fountain Life APEX members will have access to the Immunis Phase 2 clinical trial in 2025 and onward.

Insights from Viome and Naveen Jain: Measuring Gene Expression (DNA vs. RNA)

One of the technologies I've invested in and use is delivered by Viome. Viome's technology is based on a very important insight: Your "transcriptome" (which genes are on or off) is more important than your DNA sequence. While most researchers focus on your entire genome (i.e., what genes you inherited from mom and dad), as it turns out, your DNA doesn't change over the course of your life. Whether you suffer from a disease or not, your DNA is identical. What changes between being healthy and the onset and progression of the disease is your gene expression: which genes are expressed (i.e., turned on for transcription into mRNA and thereafter proteins), how they are expressed, and how the expression of these genes changes over time.

Viome's Full-Body Intelligence Test measuring transcriptomics.

So, can we measure gene expression? What if we could measure your genes and the gene expression of the nearly 100 trillion bacteria in your gut microbiome that transform food into fuel? How would these microbial genes impact your overall health? Viome uses a unique, peer-reviewed, and clinically licensed technology called metatranscriptomics to accurately measure the expression level for *all* human and *all* microbial genes in your body.

As Naveen Jain, CEO of Viome, points out, by analyzing your microbiome, adjusting your diet and nutrition, and understanding which supplements you need, you can improve your overall health while reducing your risk of disease progression.

For example, are spinach and broccoli good for everyone? It turns out that spinach and broccoli, while packed with nutrients, also contain ingredients that could be toxic to some people due to high concentrations of oxalates in spinach and sulfur in broccoli, which can be harmful if the gut microbiome is not properly processing them.

One of the outputs from Viome is an analysis, based on your gut microbiome, of what foods you should avoid and what foods are your superfoods. I don't know anywhere else to get that information. Here are the results of my last Viome Full-Body Intelligence test (just a partial listing of what I learned):

- **Foods to Avoid**: Artichoke, Black Beans, Caviar (☹), Coconut milk, Onion, Tomato...
- **Superfoods**: Almonds, Apples, Beets, Broccoli, Brussel Sprouts, Salmon, Spinach...

Viome explains why to select or avoid each food for each of these.

Viome's AI-based platform, which has been trained on millions of data points from the company's over 16,000 clinical research participants, can tell you how your unique microbiome responds to over 300 foods and how each could benefit or harm you.

What the Data Shows: Food is Medicine

Last year, Viome's scientists published (*American Journal of Lifestyle Medicine*) the results of a clinical trial that measured the effects of Viome's food and supplement recommendations and biotics on several diseases.[102] Here are some of the findings:

- 40 percent reduction in severe irritable bowel syndrome (IBS) clinical scores
- 30 percent reduction in depression among participants
- 30 percent decrease in anxiety among participants
- 30 percent decrease in diabetes risk score among participants

I've been impressed by Viome's data, research, and publications. Another validating indicator is that Viome received a "Breakthrough Device Designation" from the FDA for their RNA sequencing (metatranscriptomic) in detecting oral and throat cancer. And while Viome has already made significant strides in general health and specific conditions like depression, anxiety, IBS, diabetes, and aging, their AI is continuously learning.

Based upon Viome's Full-Body Intelligence test (transcriptomics), the company provides personalized supplements, probiotics, personalized toothpaste, and oral probiotics, which I use (as described in my dental routines at the end of Chapter 7).

Chapter 6
Longevity Mindset & Happiness Hormones

"There is a fountain of youth: it is your mind, your talents, the creativity you bring to your life, and the lives of people you love."
—Sophia Loren

What is a Longevity Mindset?

One of the most important elements of my longevity practice is creating and maintaining a longevity mindset.

A longevity mindset is one in which *you believe in the ability of science to extend your healthspan*, perhaps by ten or twenty years. Furthermore, it is a belief (and an understanding) that science isn't standing still during these additional decades. Instead, health technologies are accelerating exponentially, making breakthroughs driven by AI, genome sequencing/editing, epigenetic reprogramming, and gene therapy—all focused on extending healthspan and reversing disease.

Finally, should you understand that such health-extending technologies are on the horizon, it is a desire to keep yourself as healthy as possible to intercept these technologies when they arrive. Ultimately, a longevity mindset means becoming the "*CEO of your health*" and recognizing that "*Life is short until you extend it.*"

What Shapes Your Longevity Mindset and Your Healthspan?

How long you live is a function of many factors: where you were born, your genetics, your diet, *and* your mindset. Most people imagine that longevity is mostly inherited and that the genetic cards you are dealt have predetermined your lifespan.

As I mentioned in the introduction and am repeating here to make a crucial point: *You may be surprised by the truth.*

In 2018, scientists analyzed a 54-million-person ancestry database and announced that lifespan has little to do with genes.

In fact, *heritability is accountable for only approximately 7 percent of your longevity.*

Other studies peg this somewhat higher, estimating that heritability accounts for 20–30 percent of your lifespan—which still means, at a minimum, that *lifestyle choices account for at least 70 percent of your longevity.*

The power of shaping your healthspan is much more in your hands than you might have imagined. While we've already discussed things like diet, exercise, and sleep, one of the biggest (underestimated) impacts on your healthspan is your mindset, which we'll now explore.

Six key mindset and lifestyle areas (under your control) fundamentally impact your healthspan. Let's review each area together, and as we do, I invite you to ask yourself: "Where do I truly stand in this area? Where can I improve? What would it take for me to modify my beliefs and actions?"

1. What do you honestly believe about your healthspan?

Understanding your ingrained beliefs about your expected lifespan and healthspan is the first place to begin. At one end of the spectrum, you see life as short and precious; you'll consider yourself lucky if you make it to 75 or 80 years old. At the other end, you're focused on breaking through 100 years old with energy and passion, making "100 the new 60." In this latter mindset, you see aging as a disease that can be slowed, stopped, and perhaps even reversed. *How old do you think you'll live? Why? What are your ingrained beliefs, and where did you get them?*

Peter speaking with Abundance Platinum members, a global community of entrepreneurs and leaders that create a Virtual Blue Zone.

2. What media are you consuming?

The type of content you consume (e.g., books, blogs, movies, news) constantly shapes how you think and directly impacts your longevity mindset. Are you reading the obituaries of old friends? Or are you reading books like *Life Force* (which I co-authored with Tony Robbins and Robert Hariri, MD, PhD), David Sinclair's *Lifespan*, or Sergey Young's *Growing Young*? Obviously, you're currently reading this book, so that's a great start! Beyond books, what podcasts are you listening to? What newsfeeds do you subscribe to? Our minds are neural nets, similar to today's large language models, constantly shaped by what we see and hear. Carefully choose what you let into your mind in the same way you are careful about the foods you eat.

3. Who do you hang out with? Who makes up your community?

The people you spend time with have perhaps the biggest impact on shaping what you believe and the actions you take. Are you actively building and deeply engaging with a community that is optimistic and youthful despite its age? A group actively pursuing longevity, sharing best practices, and encouraging one another? A group that exercises together regularly? It's been said that you are the average of the five people you spend the most time with. So, who are you spending the most time with? If you are overweight and want to slim down, hang out with friends who have great eating and exercise habits.

4. Are you prioritizing sleep?

Sleep is a fundamental tool to optimize your healthspan. As mentioned in the sleep chapter, a great book that details this is *Why We Sleep* by Dr. Matt Walker. We physiologically *need* seven to eight hours of sleep per night. Do you subscribe to the motto, "There's plenty of time to sleep when I'm dead"? Or do you prioritize sleep and use the best available techniques to help you achieve eight healthy hours of sleep every night?

5. How healthy is your diet?

There is truth to the saying, "You are what you eat." Do you eat whatever you want, whenever you want? Are you overweight and eating way too much sugar? Or have you intentionally shaped your diet, minimizing sugars and high-glycemic foods while eating a diet high in whole plants and sufficient protein to build muscle?

6. How much exercise do you get?

Exercise is fundamental to longevity, along with your mindset, sufficient sleep, and a healthy diet. The latest research on longevity clarifies that maintaining and (if possible) increasing muscle mass is critical. So, where are you on the exercise spectrum? On one end, you don't exercise at all. On the other end, you're getting 10,000 steps a day and exercising with weights at least three times each week, focusing on interval and resistance training.

A Longevity Mindset Scorecard

Every year at my Longevity Platinum Trip, I ask the participants (mostly investors and family offices) to score themselves on their mindset. Measuring your mindset is the first step toward up-leveling your beliefs. I use a Mindset Scorecard (a technique borrowed from Strategic Coach°) to do this.

Take a moment to complete the scorecard below and rate yourself 1–10 in each of the five Mindset Areas. Then, add up your score (maximum is 50). Consider rating yourself on where you were before reading this book and where you hope to be a month from now.

Mindset Areas	1 — 2 — 3	4 — 5 — 6	7 — 8 — 9 — 10	Before Score	After Score
Overall Longevity Mindset (What I believe)	Life is short & precious. I hope to make it to 75.	I'm aiming to get to at least 80 yrs old, maybe to 100. My goal is to stay out of a wheelchair!	I will get past 120 yrs old (min) (healthy & fit). Shooting for 150 year healthspan or older!		
What I Read & Watch (What I let into my brain)	I don't monitor what I read or watch. Typically I'm watching CNN /Fox, sometimes reading the obituaries of old friends.	I watch some negative news. I try to read more inspiring blogs & books and educate myself about the latest health and biotech breakthroughs.	I actively shape my Longevity Mindset by consuming inspiring Books (ie: *Lifespan*) & Blogs & News (*Longevity Insider*).		
My Community (Who I hang out with)	I mostly hang out with older people who are always talking about disease, aches, pains and death.	I maximize my time with friends and loved ones, both young and old. We discuss diet, exercise and health, but not longevity.	My friends and I are young, or young-minded; We pursue longevity content and hang out with health-conscious individuals.		
My Diet, Sleep & Exercise (How I treat my body)	I eat whatever I want; I don't prioritize sleep and rarely exercise My health is not a top priority.	I know sleep, diet, and exercise are important. I'm working on prioritizing these areas and learning about best practices.	I sleep 7+ solid hours per night. I am focused on building Muscle Mass and a healthy diet with Intermittent Fasting and No Sugar.		
My knowledge of cutting-edge Diagnostics & Biotech Interventions	I have no idea what's going on inside my body. For my health, I occasionally imbibe red wine & scotch. I'm not concerned.	My health strategy involves: (i) an annual check up, (ii) statins, (iii) BP medicine & (iv) Cardiac Score. I do what my doctor orders.	I have sequenced my genome, do an annual MRI; I take Peptides, Stem Cells, Exosomes, and the latest regenerative treatments, and joined a longevity team (e.g., Fountain Life).		
			Total Scorecard		

Why Your Mindset Matters: The "Will to Live"

I want to close this Longevity Mindset Practice segment with a notion about the power of the human mind—specifically, "the will to live."

It's all about "mind over physiology."

A key mindset for longevity involves being excited about the future and having something to live for and look forward to.

Dan Sullivan of Strategic Coach®, puts it this way, "Always make your future bigger than your past."

Tony Robbins says, "Having a bigger purpose to live for is absolutely key to longevity."

My favorite story illustrating this comes from the annals of American history.

In an extraordinary demonstration of the will to live, two of America's founding fathers, Thomas Jefferson and John Adams, willed themselves to live long enough to see the 50th anniversary of the Declaration of Independence.[103]

Even though the average life expectancy was only 44 years old in the early 1800s, Jefferson (who was 83) and Adams (who was 90) made it to July 4, 1826, *both dying on that exact date*, the 50th anniversary of the nation they had founded.

Clearly, they had a goal in mind, something to live for.

So, how long do you think you'll live? Until you're 80 years old? Maybe 100?

What mindset or purpose would you require to set a target of 120 healthy years and make it there?

Your health is your greatest wealth, and today is the most extraordinary time to be alive.

Please begin to change how you discuss your lifespan (healthspan) with others. Make it known to friends and family (with conviction) that you're shooting for 100, 120, or even 156. Pick a number that inspires you and program that into your mind.

One of the most important conversations you can have with yourself, your friends, and loved ones is to ask the following question:

"What would you do with an extra 30 years of healthspan?"

Having a clear vision and emotionally connecting with *why* you want those extra decades makes all the difference in the world. *The results are powerful.*

My motivations for getting an extra 30 years of healthspan are many, but primarily the following three:

1. I had kids later in life (twin boys born when I was 50). I'm 63 now, and my 13-year-old boys are more fun than ever. One of my biggest motivations is my desire to see my children thrive and meet my grandkids and great-grandkids.

2. I'm a child of the 1960s when the Apollo program and that scientific documentary *Star Trek* showed us how far humanity can reach. I want to see humanity open the space frontier, settle the Moon, reach Mars, build O'Neil colonies, and mine asteroids. I not only want to see it all, but I also want to travel there and participate myself.

3. I'm excited to see how the area of brain-computer interfaces moves forward and how tech may enable us to connect our consciousness with the cloud and even pioneer the ability for humans to "upload" our minds. (Yes, I know this is pretty far out, but, heck, a lot can happen in the next 60-plus years!)

What are your big goals? If you're not clear about them, please take the time now to jot down some ideas or have that discussion with your loved ones. Creating a vision for your extended future is the first step to getting there.

Enabling an Extra 20 Years: $101M XPRIZE Healthspan

One of the most successful organizations I founded some 30 years ago is the XPRIZE Foundation. XPRIZE designs, funds, and launches large-scale incentive competitions to inspire breakthroughs in areas that are stuck or moving too slowly. The foundation doesn't award individuals for work done in the past, like the Nobel or Pulitzer Prizes. Instead, it sets audacious goals and challenges scientists and entrepreneurs to build and demonstrate the technology to solve key problems. We don't pay for ideas; we pay for demonstrated solutions.

XPRIZE competitions are funded by leading entrepreneurs, corporate giants, and global philanthropists like Google (funder of the $30 million Google Lunar XPRIZE), Elon Musk (funder of the $100 million Carbon Removal XPRIZE and $15 million Global Learning XPRIZE), Wendy Schmidt (funder of multiple

Peter and the XPRIZE Team raised $141 Million to launch the XPRIZE Healthspan to focus hundreds of teams on increasing our healthy lifespan.

ocean-related XPRIZEs), and Ratan Tata (funder of the Water Abundance XPRIZE).

By conducting incentive prize competitions, XPRIZE enables human capital to create scalable and transformative solutions that drive real impact.

In the last 30 years, XPRIZE has launched 30 competitions with approximately $550 million of cumulative prize purses. The competitions have yielded a 30-fold return on prize capital, driving over $5 billion in ongoing research and development. Historically, teams competing in XPRIZEs have cumulatively spent between 20 to 30 times the prize purse in their efforts to win the prize.

To illustrate the XPRIZE concept, allow me to describe the Ansari $10M XPRIZE, the first-ever XPRIZE I launched back in 1996 under the Gateway Arch in St. Louis, Missouri. This competition was designed to lower the risk and cost of going to space by incentivizing the creation of a spaceship that would finally make private space travel commercially viable. Teams from any nation were challenged to build a reliable, reusable, privately financed, crewed spaceship capable of carrying three adults to an altitude of 100 kilometers above the Earth's surface, landing safely, and making the trip again to 100 kilometers with the same spaceship within two weeks.

Twenty-six teams from seven nations competed, and on October 4, 2004, Mojave Aerospace Ventures won the $10M grand prize with the successful flights of their vehicle SpaceShipOne. Led by famed aerospace designer Burt Rutan and his company Scaled Composites, with financial backing from Microsoft billionaire Paul Allen, the team's winning technology was licensed by Richard Branson to create Virgin Galactic. With the awarding of this competition, a brand-new multi-billion private space industry was launched.

XPRIZEs work because they recruit hundreds or thousands of teams, each trying different approaches.

So, why am I writing about prizes and spaceflight in a book about longevity and healthspan? Because in late 2023, XPRIZE launched its largest global competition ever, focusing on delivering additional decades of health to us all.

On November 29, 2023, we launched the largest-ever XPRIZE competition, focused on "extending healthspan" in humans and closing the gap between life and health expectancy. This seven-year, $101 million global competition aims to catalyze the development of therapeutic treatments that restore muscle, cognition, and immune function by a minimum of 10 years, with a goal of 20 years (and perhaps more). We want to make healthy aging possible for everyone. (**Note:** There is also a $10 million bonus prize for those able to help solve a type of muscular dystrophy called FSHD (facioscapulohumeral muscular dystrophy). In success, the competition will help drive breakthroughs in the field, attracting significant capital while increasing public attention and helping accelerate the regulatory frameworks required to slow, stop, or even reverse aging. As of the writing of this book, the XPRIZE Healthspan has some 440 teams that have entered the competition.

Optimists Live Longer

In addition to the XPRIZE Healthspan, which gives me great hope that we'll have some incredible breakthroughs by 2030, I want to share a compelling piece of data on why you should be optimistic about an extended healthspan and why you should be optimistic in general.

In a study of 69,744 women and 1,429 men, published in the prestigious journal *Proceedings of the National Academy of Science*, it was found that optimistic people live as much as 15 percent longer than pessimists.[104] The study was conducted over three decades, controlling for health conditions, behaviors like diet and exercise, and other demographic information.

There is a lot to be grateful for and a lot under your control.

> "In a study of 69,744 women and 1,429 men, published in the prestigious journal *Proceedings of the National Academy of Science*, it was found that **optimistic people live as much as 15 percent longer than pessimists.**"

Happiness Hormones!

A member of my Abundance360 community, William Vipinchandre, recently published a post for members on the topic of happiness hormones. It was so beautifully written and practical in its presentation that I'm proud to share it below with minor edits.

Achieving longevity and well-being involves nurturing our body's natural processes, including the production of happiness hormones (neurotransmitters). These hormones—dopamine, endorphins, oxytocin, and serotonin—play crucial roles in our emotional and physical health. By understanding and optimizing them, we can lead happier, healthier, and longer lives.

Here's how each hormone works and how to boost them effectively:

Dopamine (For Pleasure + Reward)—Dopamine is the hormone responsible for pleasure and reward. It motivates us to pursue activities that bring us joy and satisfaction. To optimize dopamine, consider:

- Eating Food 🍎: Enjoying a nutritious meal can stimulate dopamine release. Focus on a balanced diet with plenty of fruits, vegetables, and whole grains.
- Achieving a Goal 🎯: Setting and accomplishing small, manageable goals can boost dopamine levels, providing a sense of accomplishment.
- Completing a Task ✅: Even simple tasks like tidying up or organizing your workspace can trigger dopamine release.
- Self-Care Activities 🛁: Regular self-care practices, such as taking a relaxing bath or enjoying a hobby, can enhance dopamine production.

Endorphins (For Pain + Stress) 💪—Endorphins act as natural painkillers and stress relievers. They help us feel good and reduce discomfort. To increase endorphins, explore:

- Laughter 😄: Engage in activities that make you laugh, such as watching a comedy show or spending time with funny friends.
- Exercising 🏃: Regular physical activity, especially aerobic exercise like running or cycling, is a powerful way to boost endorphin levels.
- Listening to Music 🎵: Enjoying your favorite music can elevate your mood and trigger endorphin release.
- Watching a Movie 🎬: Watching a film you enjoy, whether it's a comedy or an inspiring story, can also help release endorphins.

Oxytocin (For Love) 💜—Oxytocin, often called the "love hormone," is crucial for social bonding and emotional connections. To enhance oxytocin, try:

- Socializing 👥: Spend quality time with loved ones, whether it's through a shared meal, group activity, or meaningful conversation.
- Physical Touch 🤗: Hugs, handshakes, or any form of gentle physical contact can stimulate oxytocin production.
- Petting Animals 🐾: Interacting with pets, such as stroking a dog or cat, can increase oxytocin levels and reduce stress.
- Helping Others 🤝: Acts of kindness and helping others can significantly boost oxytocin, promoting a sense of connection and well-being.

Serotonin (For Good Mood) 😊—Serotonin is essential for mood regulation, promoting happiness and well-being. To optimize serotonin, use:

- Meditation 🧘: Practicing meditation and mindfulness can improve serotonin levels and enhance overall mental health.
- Sun Exposure ☀️: Spending time in sunlight helps the body produce Vitamin D, which is linked to serotonin production. Aim for at least 15-30 minutes of sun exposure daily.
- Being with Nature 🌳: Immersing yourself in natural environments, such as walking in a park or hiking in the woods, can boost serotonin.
- Mindfulness 🧘: Engage in activities that promote mindfulness, such as yoga or deep-breathing exercises, to increase serotonin levels and reduce stress.

Integrating Happiness Hormones for Longevity ✨—By consciously engaging in activities that boost dopamine, endorphins, oxytocin, and serotonin, we can create a balanced and more fulfilling lifestyle that promotes longevity. These hormones enhance our mood, reduce stress, and improve our overall physical health. Incorporate these practices into your daily routine to build a life filled with joy, connection, and vitality, paving the way for a longer, healthier existence.

Embrace the power of your happy hormones and let them guide you toward a life of lasting wellness and happiness.

William Vipinchandre recommends Serenity Haven's wellness retreats 🌿 as a solution designed to profoundly nurture your body and soul through holistic practices, helping you optimize your happy hormones and foster a life of longevity, health, and happiness. More information can be found at SerenityHaven.io.

Chapter 7
Winning Habits: Routines That Keep You Young

*"We are what we repeatedly do.
Excellence, then, is not an act but a habit."*
—**Aristotle**

*"It's not what we do once in
a while that shapes our lives.
It's what we do consistently."*
—**Tony Robbins**

One of the biggest changes I made in my health isn't the result of a supplement or workout practice but the result of instituting daily routines—stuff I do every day, the habits that provide me consistency and make healthy actions second nature.

In this *Longevity Guidebook*, I've included an entire chapter on routines because I believe routines are one of the most important mechanisms of creating long-lasting positive changes to your health. The following pages outline what I try to accomplish every day and why. Realistically, in a good week, this will happen five or six days out of seven, and yes, this will get seriously disrupted when I'm traveling, but I still do my best.

As Kevin Kelly (founder of *WIRED* magazine) said about routines and habits, "The purpose of a habit is to remove that action from self-negotiation." You create a habit of automating a process. You don't have to decide if you perform a habit enough times. Deciding causes decision fatigue, which sucks away your precious daily energy.

Chances are that you already have some type of morning routine, and the question is, does it fully serve your health objectives? I hope the following sections will inspire you to update and optimize what you do every day and every morning. Borrow from this chapter what you desire and customize it to serve you best.

"Win the Morning to Win the Day."

"Early to bed and early to rise makes a man healthy, wealthy, and wise." —Benjamin Franklin

In my 30s and 40s, I never thought about morning routines. Now, I think it's one of the most important things I do.

Creating a consistent morning routine is a great way of implementing health-related habits. The early morning hours are a fresh start, a clean slate, and typically the part of your day with maximum willpower and (for me) maximum freedom of schedule.

Throughout medical and graduate school, I was a night owl, staying awake till 2:00 a.m., minimizing sleep, and powering through the day on five cups of coffee. That's changed a lot. Over the last decade, my schedule has shifted to what I view as a much healthier and far more productive routine.

So, here's my morning routine for your consideration, specifically what I do and why I do it. The days when I'm most productive, most fulfilled, and happiest are when I can follow my routine. My routine sets me up for maximum success, and I wish the same for you.

Waking Up: 5:30–6:00 a.m. (First 30 Minutes)

My morning routine begins the night before by getting to sleep on schedule, typically by 9:30 p.m. Getting eight hours of sleep is a priority. When I'm in a regular rhythm, I wake up on my own, typically at 5:30 a.m., and while I set an alarm as a backup, it rarely wakes me.

Here's what I do as soon as I wake up:

- **Mindset:** First, I get excited about the day ahead and give thanks for life. I view life as amazing, truly a miracle. If you're in reasonably good health and can afford this book, you're one of the lucky people on Earth. Kicking off the day with a mindset of gratitude and optimism is essential. A fresh morning start allows us to let go of yesterday's frustrations and embrace the new day of near-infinite possibilities.

- **Check My Oura Score:** Since I care (a lot) about sleep and have set a goal of eight hours of sleep, with seven hours as a bare minimum, checking my Oura Ring score first thing helps me measure and gamify my sleep. Ultimately, my objective is a Sleep Score and Readiness Score of 90 or more. Hitting that target motivates me to get to bed by 9:30 p.m.

- **Oral Routine:** I'm focusing on oral healthcare more than ever; it drives brain, cardiac, and microbiome health (see the upcoming two-page box for more details). Here's what I do every morning and night.

    - **Proclaim Custom-Jet Oral Health System**: It's an amazing 3D-printed mouthpiece called Proclaim that shoots individual jets of water (like a Waterpik) between all my teeth. It only takes 15 seconds—powerful and fast!
    - **Hydrosonic Electric Toothbrush:** I use this with a custom toothpaste designed for my oral microbiome provided by Viome following Viome's oral microbiome testing.
    - **Dental Herb Co.'s Tooth and Gums Tonic Mouthwash**

- **Skin Routine:** I use OneSkin's OS-O1 face cream daily (and their body lotion) to kill senescent skin cells and reverse skin aging (see the chapter on medications and supplements for more details).
- **Hydration:** Drinking water (typically two cups) as soon as possible after waking up helps rehydrate the body and improve digestion. Adding lemon juice or apple cider vinegar to the water can stimulate digestion and maintain gut health.
- **A Cup of Coffee, Matcha Tea, or Nutri11:** I limit myself to a *single cup* of caffeinated coffee. Any other coffee I might have during the day is decaffeinated. Occasionally, I'll substitute green tea, matcha green tea, or a hot mug of Nutri11.
- **Natural Light Exposure**: Within the first hour of waking up, whenever possible, I try to get a few minutes of exposure to natural sunlight (even through cloud cover). This helps me regulate my body's internal clock and improve overall mood and energy levels by increasing my early-day cortisol release. This is most valuable when I travel overseas and want to reset my internal clock. A morning spike in cortisol will also positively influence your immune system, metabolism, and ability to focus during the day.[105]

6:00–6:30 a.m. (Redlight and Meditation)

My next 30 minutes involve "stacking" a number of elements together for the efficiency of time and maximum value, meaning I'm doing many of these things simultaneously.

- **Red Light Therapy (Body, Scalp, and Mouth):** I expose my body to 20 minutes of red light therapy *simultaneously from three devices all stacked together*:

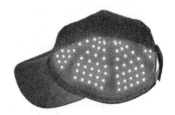

Kiierr laser hair growth cap.

  1. I use a **PlatinumLED Biomax 900** red light panel to reduce inflammation and promote healthy skin (20 minutes).
  2. At the same time, I use a **Kiierr laser hair growth cap** to stimulate hair growth. I wear the cap, lined with two hundred red laser

diodes, for 20 minutes (used every other day for 20 minutes per session).

3. Finally, every day, for about 5 minutes, I use a **Guardian+ BioLight mouthpiece** that bathes my teeth and gums with red light therapy to kill harmful bacteria in my mouth and support gum health (5 minutes).

- **Vagal-Nerve Stimulation:** During my red light session, I put on a Pulsetto "vagal-nerve stimulator" around my neck. It activates the vagus nerve by increasing heart rate variability (HRV), lowering blood pressure and stress, reducing inflammation, and improving mood. I will typically repeat this at some point later during the day. I keep my Pulsetto on my desk within easy reach and eye-sight to remind me (5 minutes).

Peter uses his Pulsetto vagal nerve stimulator at least twice per day to put him into parasympathetic model and increase his HRV.

- **Meditation:** Finally, during this window, I'll do a 15–20-minute meditation every morning, typically using my MUSE device, which has some great meditations, as well as brainwave monitoring to guide the meditation. **Note:** I can *not* use the MUSE at the same time as my Pulsetto because of electrical interference, so I use them back to back.

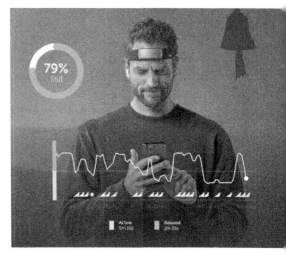

Peter uses a Muse device to support his morning meditations.

6:30–7:30 a.m. (My "Golden Hour")

This next hour is my most productive, clear-minded "golden hour." I'm typically most focused and get my best writing done during this part of the day. Here's what I focus on next:

- **Process My Email:** First, I'll review the emails that came in overnight. Touching each email once: delete, forward, or answer them accordingly. Email can be a time suck, so I do my best to limit my email time to 10 minutes just to identify urgent matters that might make it to my top-five list or might cause me to rearrange my schedule (10 minutes).

- **Prioritize My Day—My Top 5:** One of the most important things I do at the start of the day is to look at my schedule and examine the laundry list of *everything* I'm thinking about achieving during the day ahead. Next, I prioritize my top five items. I keep a written "Action Items" list of everything, what I did and didn't get done yesterday, projects I'm thinking about, and people to follow up with. After prioritizing my top five, I mentally note, "If I do these, then today is a huge win! *Everything else is a bonus*" (5 minutes).

- **Writing:** I love writing. Whether it's one of my blogs or a book like this one. I try to focus on writing for at least 45 minutes every morning. I have a set of music selections stored as "my music" on Amazon Alexa that helps put me in a focused state. Sometimes, I play 100 Hz binaural beats off YouTube to block noise and keep me focused. You can check out what I listen to by searching on Google for "100 Hertz Super Intelligence Memory Music."

7:30–8:30 a.m. (Workout)

> *"We do not stop exercising because we grow old; we grow old because we stop exercising."*
> —Kenneth Cooper, MD

My goal is to exercise seven days per week. I'm clear that exercise is my number-one pro-longevity protocol.

Regarding exercise, I've split my days into two types: (1) Weight Lifting/Muscle Gain days and (2) VO2 Max/Fasting days. Allow me to briefly explain below, but please refer to the chapter on exercise for more details.

Weight Lifting/Muscle Gain Days

On these days, my goal is to lift to gain and maintain muscle mass. These days are typically Wednesday, Thursday, Saturday, and Sunday. My routine for those days involves:

- **Protein Shake:** Before I work out, I prepare a protein shake, typically alternating between Ka'Chava (plant-based) or a high-quality Whey protein shake. The shakes have about 30 grams of protein. I include 5 grams of creatine in the shake and a Mitopure supplement that contains urolithin A (shown to stimulate a process called mitophagy, which is the cell's way of cleaning out old, damaged mitochondria and replacing them with new, healthier ones), my probiotic powder supplied by Viome, and half an avocado and/or frozen blueberries.
- **Stretching/Daily Mobility:** During the past few months, I've added a stretching routine to the beginning and end of my workouts after my muscles have warmed up. I have developed a healthy respect for the critical nature of these exercises and do them, if possible, seven days per week.
- **Weight/Resistance Workout:** I do short but intense 40-minute weight workouts targeting my upper and lower body. Where possible, I am using a trainer or my Tonal (AI-enabled) system to make sure I'm doing the exercises properly and avoiding the possibility of injury.
- **Additional Protein**: After my workout, I typically consume additional protein in the form of two or three cooked egg whites to support muscle growth.

VO2 Max/Fasting Days

On these days, my goal is to focus on interval training to stimulate my mitochondria and improve my VO2 max. These typically fall on Monday, Tuesday, and Friday, though I will try to squeeze in 15 minutes of HIIT training (see below) on my bike on other days as well.

On these days, when I'm not lifting weights, I focus instead on intermittent fasting and delay my eating until the afternoon, typically eating between 2:00–7:00 p.m.

Fasting: Fasting involves limiting myself to coffee, green tea, or water from waking through 2:00 p.m., followed by dinner around 5:30 p.m. My goal is to limit my eating to five hours. During my fasting period (7:00 p.m.–2:00 p.m.) of 19 hours, I keep my stomach full with water to help minimize any hunger pangs. An unexpected consequence of this fast is I actually have *more energy* because blood flow is not diverted to my digestive system.

HIIT Training/VO2 Max: My goal is high-intensity interval training (HIIT) using my Technogym stationary bike. I warm up for 5 minutes and then alternate between 1 minute of high-intensity exercise (fast cycling) and 1 minute of low-effort peddling for recovery. See the chapter on exercise for more detail (15 minutes).

Famous Morning Routines

I find morning routines a fascinating subject. Below are three other routines for your review and consideration.

Steve Jobs: Daily Self-Reflection as a Compass for Life

Steve Jobs, the visionary co-founder of Apple, was known for his intense focus on what truly mattered. Every morning, he engaged in a simple yet profound routine that set the tone for his day. Standing in front of the mirror, Jobs would look himself in the eye and ask, "If today were the last day of my life, would I want to do what I am about to do today?" This question wasn't just rhetorical; it was a powerful tool for self-reflection that helped him evaluate whether his daily actions aligned with his deepest values and long-term goals. When the answer was "no" for too many consecutive days, Jobs took it as a clear signal that something in his life needed to change. This daily practice of intentional questioning served as a personal compass, ensuring he remained on a path of purposeful living, and it can inspire us to regularly assess our priorities.

Benjamin Franklin: Morning Journaling for Moral and Personal Growth

Journaling has long been recognized for its myriad benefits, from reducing stress to boosting productivity. Benjamin Franklin, one of the United States' founding fathers, strongly advocated this habit. Starting at the age of 20, Franklin used his journal not just as a tool for recording daily events but as a means of self-improvement. Each morning, he would reflect on the question, "What good shall I do this day?" Intentional goal-setting and moral contemplation were central to Franklin's lifelong pursuit of virtue. Nearly 300 years later, his approach to journaling remains a powerful model for anyone seeking to enhance their character and start the day with a clear sense of purpose. By incorporating even a brief moment of reflective journaling into our mornings, we can better align our daily actions with our broader life goals.

Tim Cook: The Power of Waking Up Early

Tim Cook, the CEO of Apple, exemplifies the discipline of early rising, a habit that many successful leaders share. Cook's day begins at 3:45 a.m. when most of the world is still asleep. This early start allows him to address critical tasks with undivided focus and work out before the demands of the day take over. Research supports the idea that early risers tend to procrastinate less and are often more productive than night owls.[106] While waking up before dawn might not be feasible for everyone, the essence of this habit lies in its consistency. By gradually shifting your wake-up time, even by just 10 minutes at a time, you can create a morning routine that sets a productive tone for the entire day. Whether it's for personal reflection, exercise, or quiet work, carving out those early hours can be transformative.

My Dental Protocol

Why Oral Health Matters to Longevity

Oral health plays a crucial role in overall well-being, and it's often overlooked when considering the root causes of various health issues. The connection between oral health and systemic diseases is well-documented and profound. Poor oral health is not just about cavities and gum disease; it can also be a precursor to serious health problems such as heart disease, dementia, autoimmune disorders, and cancer.

According to the CDC, nearly 50 percent of adults aged 30 and older have some form of periodontal disease.[107] This prevalence increases with age, with 70 percent of adults aged 65 and older being affected by periodontal disease.

Individuals with gum disease are about 20–25 percent more likely to develop heart disease.[108] This connection is due to the inflammation caused by oral bacteria entering the bloodstream, contributing to arterial plaque formation.

Oral bacteria or inflammatory cytokines from gum disease may enter the bloodstream and cross the blood-brain barrier, contributing to the formation of amyloid plaques and neurofibrillary tangles in the brain, which are associated with Alzheimer's disease.

Porphyromonas Gingivalis and Alzheimer's Disease

Research has identified the oral bacteria Porphyromonas gingivalis, which is commonly associated with chronic gum disease, in the brains of patients with Alzheimer's disease. A 2019 study published in the journal *Science Advances* found that *P. gingivalis* and its toxic enzymes, known as gingipains, were present in the brains of Alzheimer's patients.[109] This suggests that oral bacteria might contribute to the development of neurodegenerative diseases by entering the bloodstream and reaching the brain. A 2020 study published in the *Journal of the American Geriatrics Society* found that older adults with severe periodontitis had a 20–30 percent increased risk of developing dementia compared to those with healthy gums.[110]

My Oral Health Protocol

It's for all the reasons above that I've "upped my game" regarding oral health over the past few years. Here's an expanded view of my dental-oral protocols.

Hydrosonic Electric Toothbrush (by Curaprox): Brushing twice daily is still the basics. Historically, I've used a toothpaste called OnGuard, but recently, I switched to a toothpaste provided by Viome, customized for my oral microbiome, called a Sunrise Cleanse & Restore Gel (morning) and a Nightly Nourish & Protect Toothpaste (evening).

Redlight Mouthpiece: I use a guardian redlight mouthpiece by Biolight. There are numerous options on the market. Redlight is thought to reduce gum inflammation and eliminate harmful bacteria.

Proclaim Custom-Jet Oral Health System: I've always struggled with a Waterpik device, trying to get the jet into the right place and doing it without making a mess in the bathroom. So recently, I was thrilled to learn about and begin using an amazing 3D-printed mouthpiece called Proclaim that shoots individual jets of water (like a Waterpik) between all my teeth. Takes 15 seconds. Powerful & fast!

Flossing: Yes, flossing is important. My hack is to floss during my drive time. So, I carry around a bunch of Plackers Twin-Line Dental Flossers in the center console of my car and floss a couple of times per day while in LA traffic.

Chewing Xylitol Gum: I recently learned that chewing sugar-free xylitol gum is actually very healthy for our teeth. Xylitol is a naturally occurring sugar alcohol widely used as a sugar substitute in sugar-free products and has 40 percent fewer calories than sugar. Xylitol is known to reduce the levels of harmful bacteria in the mouth, particularly *Streptococcus mutans*, which is a major contributor to tooth decay.[111] Xylitol gum also helps neutralize mouth acids produced by bacteria after eating, maintaining a healthier pH level, which is less conducive to tooth decay.

Biolight redlight mouth piece for gum health.

Dental Herb Tooth & Gums Tonic Antimicrobial Mouthwash: I use this Tooth & Gums Tonic® after brushing, morning and night. It's an all-natural mouth disinfectant (essential oils and herbs) to reduce harmful bacteria and gum inflammation.[112]

Oral Probiotics: Every night after brushing, I take a probiotic lozenge from Viome tailored to my oral microbiome. My custom formulation contains Lactobacillus plantarum L-137, lycopene, Streptococcus salivarius K12, xylitol, inulin, L-leucine, dicalcium phosphate, Ligilactobacillus salivarius, Lacticaseibacillus casei, Bifidobacterium bifidum, and Lacticaseibacillus paracasei.

Oral Dental Appliance (Mandibular Advancement Device): As discussed in the sleep chapter, I use a specially fitted upper and lower mouth guard called a "mandibular advancement device." This device is a dental appliance used to treat sleep apnea and snoring. The appliance also prevents me from grinding my teeth. I love it so much that I can't sleep without it. There's no single brand I can recommend. You'll need to visit your dentist to get upper and lower impressions taken to get it fitted and manufactured.

Deep Cleaning: I get a deep cleaning on my teeth at the dentist's office every quarter.

It should go without saying that if you smoke, quit. Smoking weakens your immune system, making it harder for your body to fight gum infections. It also increases the risk of oral cancer, periodontal disease, and tooth loss. *Smokers are four times more likely to develop gum disease*, and twice as likely to lose teeth than non-smokers.[113]

Insights from Arianna Huffington and Thrive Global

Do you need help making small but meaningful changes to build healthier habits? Enter Thrive Global. After experiencing a health scare brought on by exhaustion, Arianna Huffington launched Thrive Global in 2016 with a bold mission: to use behavior change technology to improve health outcomes and productivity. Arianna founded Thrive Global to tackle the modern-day burnout crisis and stress-induced health problems by promoting healthier habits and greater well-being. Thrive Global combines cutting-edge AI-based technology with practical well-being strategies to empower individuals and companies to achieve sustainable changes in behavior.

At its core, Thrive Global focuses on fostering a culture shift by promoting microsteps—small, science-backed changes that lead to healthier and more productive lifestyles. These microsteps are built into Thrive's digital platforms and content, guiding people through actionable steps across the five behaviors that govern our health: sleep, food, movement, stress management, and connection. The company's initiatives range from digital and live coaching programs to employee well-being platforms and leadership webinars, which companies like Accenture, Salesforce, and Walmart have adopted to help employees combat burnout and improve work-life integration. The underlying idea is that small behavioral changes, when compounded over time, can profoundly affect overall health and well-being.

Thrive Global has seen significant success in scaling its mission to a global audience. Through partnerships with major corporations, Thrive has reached millions of employees, providing them with the tools to break the cycle of stress and burnout. It boasts an impressive client list that includes Fortune 500 companies, healthcare providers, and educational institutions. The company reports that 97 percent of program participants feel better equipped to manage stress, while 95 percent have improved their overall well-being. These results underscore Thrive Global's effectiveness in improving individual well-being and driving tangible business outcomes, solidifying its position as a leader in the corporate well-being space.

Millions of daily interactions with Thrive show that focusing on well-being drives increases in employee productivity, engagement, and energy. The numbers quoted by the company are impressive:

+16 percent increase in energy levels
+62 percent increase in employee mental health scores
+37 percent increase in focus scores
+25 percent increase in resilience scores
+59 percent increase in stress management scores

In 2024, eight years after its founding, Arianna teamed up with OpenAI CEO Sam Altman to launch an expanded version called *Thrive AI Health*, which is building a "hyper-personalized" AI health coach available through a mobile app and Thrive Global products.

Huffington says she's optimistic about AI's potential to be a "key to behavior change" by assimilating large data sets, recognizing patterns, and making real-time personalized recommendations. "It's not enough to give people a generic 10,000 step goal or suggest the Mediterranean diet," says Huffington. "We now have an opportunity to use the miracle drug of behavior change to both prevent disease and optimize the treatment of disease. We need to democratize what people in the 1 percent know and practice."

The platform's AI health coach will give personalized sleep, food, fitness, stress management, and social connection recommendations. Users can give the coach as much health information as they want—not only lab, medical, and biometric data but also which foods they love and don't love; how and when they're most likely to walk, move, and stretch; and the most effective ways they can reduce stress.

Huffington concludes: "We would never scale enough 1:1 human coaches to reach everyone who needs it. AI gives us the scale, precision, and hyper-personalization possibilities to power the transformation of health outcomes." For more information, check out **ThriveGlobal.com**.

Chapter 8
Women's Health: Longevity Through Every Stage of Life

"Women's longevity is not just about living longer; it's about living better and healthier."
—Vivian Pinn, MD

WOMEN'S HEALTH

By Helen Messier, PhD, MD
Chief Medical and Scientific Officer, Fountain Life
and **Mona Ezzat-Velinov, MD,** *Apex-Extended Physician, Fountain Life*

Helen Messier, PhD, MD, Chief Medical Officer *Mona Ezzat-Velinov, MD*

Note 1: This chapter is written by two amazing women physicians at Fountain Life, with whom I have the pleasure and honor to work. This chapter is written in a singular voice, "I" vs. "we," but reflects the input of both Drs. Messier and Ezzat-Velinov, their joint experience and perspective.

Note 2: If you are a man reading this book, I encourage you to read this chapter and not skip it. It's important for us to deeply understand what our spouse, girlfriend, mother, sister, and friends are experiencing and how to support them.

Cultivating a Longevity Mindset for Women

A longevity mindset means nurturing the whole of you—today and for the future. As women, we may navigate many life stages: education, career, relationships, pregnancy, parenting, and often caregiving for both older and younger generations. Every woman's journey is unique, but we share a common thread in the experiences of working, caring, exploring, and pursuing happiness.

Thinking about longevity, even in your 20s, is invaluable. Longevity is not just something in the distant future; it's about vibrancy in your everyday life. By adopting health practices early, you can support your long-term well-being.

At the heart of longevity lies a comprehensive approach to health. Understanding that longevity practices must be personalized—especially for women—is key to optimizing your health at every life stage. As a physician, I evaluate key aspects of your body's function: gut microbiome, hormones (including stress hormones), cardiovascular health, detoxification processes, immune system function, and essential factors like sleep, movement, nutrition, and circadian rhythm. To further understand your health, we use imaging techniques like whole-body MRI to detect early signs of disease, cardiac scanning for cardiovascular risk, and DEXA scans for bone and muscle health.

Why Women Need Their Own Longevity Practices

A woman's physiology is fundamentally different from that of a man. As Dr. Stacy Sims aptly stated, "Women are not small men." Our hearts and blood vessels are smaller, we have different hormonal systems, and our bones are less dense. Our immune responses also vary, making us more prone to autoimmune diseases but less likely to develop certain cancers compared to men.

These differences demand that women adopt distinct health practices for longevity. For example, men have a 50 percent chance of developing cancer, while the rate is only 33.3 percent for women. However, women have higher rates of inflammation, autoimmune diseases, Alzheimer's, and stroke, which require a different approach to prevention and health management.

Despite these clear physiological differences, research into women's health is still limited. Women were excluded from drug trials until 1977, and even now, clinical studies and basic science research rarely analyze results by sex. For instance, while it is known that hormones like estrogen have profound effects on the brain, less than 0.5 percent of neuroimaging literature accounts for this. Similarly, only 10 percent of immune studies consider sex differences despite their critical role in immune function. Recognizing and addressing these gaps is empowering for women. The more we understand our unique physiological makeup, the better equipped we are to extend our healthspan.

Hormones

The Importance of a Holistic Approach to Hormone Balance and Replacement

Every woman is familiar with the varying effects of hormones—feeling energetic and powerful on some days and dealing with mood swings, fatigue, or weight gain on others. These changes are often signs of hormone imbalances.

Balanced hormones are crucial for overall health. Estrogen and progesterone, for instance, not only support reproduction but also maintain the health of our brain, bones, heart, blood vessels, skin, hair, and more.[114] In fact, estrogen receptors are present in nearly every cell in the body. Research even shows that the brain remodels itself in sync with the menstrual cycle, with high estrogen and low progesterone levels mid-cycle, enhancing memory and spatial cognition.[115]

When these hormones rapidly decline during menopause, many of us experience hot flashes, night sweats, vaginal dryness, weight gain, and brain fog. Sleep disturbances and mood changes are also common. Beyond these symptoms, menopause increases the risk of conditions like osteoporosis, cardiovascular disease, dementia, and some cancers.

With ongoing discussions around hormone replacement therapy (HRT) for perimenopausal and postmenopausal women, many wonder: *What are optimal estrogen levels? Which HRT approach is best?* The answers are more nuanced than simply measuring hormone levels.

Optimal hormone management requires understanding how our bodies naturally produce and cycle hormones, how hormone receptors work, and how hormones are metabolized. This comprehensive understanding is key to achieving hormonal balance.

Understanding How Hormones Are Metabolized

Hormone metabolism is a complex yet beautifully coordinated process that responds to your body's needs. Whether hormones are naturally produced or taken as therapy, they are metabolized in the liver and kidneys and eventually excreted through stool and urine. They can also be utilized in other ways. For example, during periods of stress, the body diverts resources toward producing cortisol, the stress hormone, which can reduce the production of progesterone or estrogen. This is one reason why women may stop menstruating when under high stress. This shift reflects how dynamic and responsive hormone metabolism can be.

Hormone Production and Types of Estrogen

Hormone production begins with a key building block: pregnenolone, derived from cholesterol. Pregnenolone is essential for creating other hormones like DHEA and testosterone, both of which play important roles in women's health. Testosterone, for instance, isn't just important for men; women also need it, and it acts as a precursor for estrogen production.

It's also important to note that there are three main types of estrogen, each with unique roles:

1. **Estrone (E1)**: A weaker form of estrogen, primarily present after menopause. It is mainly produced in fat tissue.
2. **Estradiol (E2)**: The most potent estrogen, dominant during a woman's reproductive years. It regulates the menstrual cycle and supports bone, heart, and brain health.
3. **Estriol (E3)**: A weak estrogen associated with pregnancy, produced by the placenta. Outside of pregnancy, estriol levels are low.

These estrogens bind to different receptors with varying strengths, triggering diverse effects. Once their job is done, they are metabolized and excreted.

Pathways of Estrogen Metabolism

The way your body processes estrogen depends on factors like genetics and nutrition. Estrogens can be metabolized along different pathways, leading to either harmless metabolites or harmful ones that may cause DNA damage. For instance, the 4-hydroxy estrogen metabolite is associated with increased breast cancer risk.[116]

Key enzymes involved in estrogen metabolism include:

- **Cytochrome P450 enzymes (CYP1A1, CYP1B1, CYP19A1)**, which influence how estrogens are processed, and
- **COMT** is an enzyme that uses methyl groups to neutralize harmful metabolites.

Genetic variations can affect the efficiency of these enzymes, influencing how well you metabolize hormones. Nutrients like sulforaphane (from broccoli) or curcumin (from turmeric) can support these pathways. Other compounds, like DIM and I3C from cruciferous vegetables, may even lower overall estrogen levels.

The Gut Microbiome's Role in Hormone Balance

Your gut microbiome, the community of bacteria and other microorganisms living in your digestive tract, plays a crucial role in regulating estrogen levels. The estrobolome, a collection of gut bacteria and fungi that metabolize estrogen, influences how much circulating estrogen remains in your system.

An imbalance in gut bacteria (known as dysbiosis) can lead to higher levels of β-glucuronidase, an enzyme that breaks down the tags your liver uses to eliminate estrogen. When this happens, estrogen can be reabsorbed into your bloodstream, potentially leading to hormonal imbalances. Maintaining a healthy, diverse gut microbiome with a diet rich in fiber and plant-based nutrients can support proper estrogen metabolism.

Progesterone and Sleep

Progesterone, another key hormone for women, has additional benefits beyond reproduction. When metabolized, progesterone produces allopregnanolone, which can bind to GABA receptors in the brain, promoting relaxation and helping with sleep. This is why supplementing progesterone can be especially helpful for women experiencing sleep disturbances, especially during menopause.[117]

Hormones and Longevity: The Importance of Balance

As all women have experienced, hormones are not only essential to how we feel in the moment but also for our overall health and longevity, especially as they decline during our menopausal years.

This is where HRT comes into play. It's important to realize that hormone management is more complex than simply replacing estrogen or progesterone. Achieving optimal balance involves understanding how our bodies naturally produce and metabolize hormones. By working with healthcare professionals, we can ensure our hormone levels stay balanced, supporting long-term health.

What Does Optimal Hormone Balance Look Like?

During our healthy premenopausal states, hormone levels are not static; they fluctuate throughout the month. At the beginning of the month, estrogen and progesterone levels are low. Estrogen peaks mid-month (during ovulation), then drops sharply. As estrogen decreases, progesterone rises, supporting the build-up and shedding of the uterine lining, resulting in menstruation.

This intricate process is controlled by signals from the hypothalamus in the brain to the pituitary gland, which in turn directs the ovaries to produce estrogen and progesterone.

Hormone Replacement Therapy and Longevity

HRT is a powerful tool for supporting longevity. When tailored to your individual needs, HRT can restore optimal hormone levels, improve how we feel, and help protect against age-related diseases like osteoporosis, cognitive decline, and cardiovascular disease.

When considering hormone replacement, it's beneficial to mimic the natural cycle as closely as possible. If we provide a constant dose of estrogen and progesterone every day, the body's hormone receptors may downregulate, reducing the effectiveness of the hormones over time. Instead, hormone replacement strategies that mirror natural hormone fluctuations tend to support better long-term health.[118][119]

By working with your doctor to assess your hormone metabolism, genetics, and gut health, you can personalize your approach to hormone balance. Whether through medications, herbal supplements, or lifestyle changes, maintaining hormonal balance is key to enhancing your well-being and longevity.

WOMEN'S HEALTH

My Approach to Hormone Replacement

Types of HRT

- Implants
- Sprays
- Vaginal rings & suppositories
- Creams & gels
- Pills & tablets
- Skin patches

There are many different types of HRT to explore.

When prescribing hormones, it's important to focus on physiological hormone replacement, designed to closely mimic a natural menstrual cycle. This involves fluctuating hormone levels followed by a withdrawal bleed, which reflects the body's natural processes. I avoid continuous combined hormone therapy, where estrogen and progesterone are given at the same dose daily without cycling. Although this approach is popular due to its simplicity and lack of bleeding, I've found that it often loses effectiveness over time. The likely cause is receptor downregulation, meaning the hormone receptors become less responsive.

In many cases, I don't find it necessary to prescribe testosterone replacement in addition to estrogen and progesterone. The body's natural production from the adrenal glands, coupled with receptor upregulation, is sufficient for many women. In some cases, though, testosterone supplementation may still be needed.

The physiological approach to hormone replacement is more involved, requiring two creams twice a day with dosage adjustments depending on the day of the month. However, many consider the payoff worth it. After about three months of consistent use, many women start to notice increased energy, vitality, brain function, and other positive changes. Blood tests should be used to confirm proper hormone levels, but the improvements in how a woman feels are often the best indicator.

An alternative option is using a bioidentical estrogen patch with oral progesterone added cyclically throughout the month. This method also supports receptor balance and can be especially beneficial for improving sleep and helpful in managing stress. Ultimately, the key is finding the right combination that works best for each individual.

Remember that lifestyle factors like sleep, exercise, nutrition, and stress management all support your hormonal health.

Longevity, Hormones, and Health

When hormone replacement therapy is done thoughtfully and in alignment with the body's natural rhythms, it can be an incredibly effective way to support women's longevity. Hormones play a vital role in everything from bone health to mental clarity. By working with your healthcare providers to monitor and personalize your treatment plan, you can ensure you're living your healthiest and most vital life.

Balancing hormones is just one piece of the puzzle. Gut health, stress management, diet, and exercise contribute to a woman's overall hormonal health and long-term vitality. The more we understand and care for our bodies, the better equipped we are to age gracefully and healthfully.

Insulin and Leptin

While we've primarily discussed sex hormones in relation to longevity, it's essential to recognize the importance of maintaining balance across all hormones, including insulin and leptin. Achieving stable blood sugar levels helps prevent diabetes and supports long-term health.

Insulin plays a key role in this balance. Your diet significantly impacts insulin levels, so it's important to avoid sugar, artificial sweeteners, processed foods, and alcohol. Insulin production follows a circadian rhythm, with levels naturally decreasing later in the day. Eating earlier, making lunch your main meal, and incorporating a 10-minute walk after each meal are effective strategies for maintaining optimal insulin and blood sugar balance.

If you've been diagnosed with gestational diabetes during pregnancy, you're at a higher risk of developing type 2 diabetes later in life. Therefore, monitoring insulin and blood sugar levels early is crucial. Keep in mind that insulin rises before blood sugar levels increase, making it important to check fasting insulin levels regularly.

Leptin is the satiety hormone produced by both adipose tissue and the small intestine. It helps regulate energy balance by suppressing hunger and promoting fat reduction in adipocytes. Leptin resistance occurs when the body no longer responds to the hormone's signals of fullness, which can lead to overeating and weight gain. However, leptin affects more than just appetite and weight. An article published in the *Journal of Medicine and Life* highlights leptin's broader influence on the immune system, cancer risk, blood pressure, fertility, and hormone regulation.[120] Testing fasting leptin levels in the blood can provide insight into your body's hormonal balance and overall health.

To effectively manage blood sugar levels and maintain overall health, it's important to understand what influences your insulin levels. Using a continuous glucose monitor can help track fluctuations. Focus on structured meals instead of snacking and prioritize high-quality protein sources. Consider incorporating supplements like berberine (500 mg, 2-3 times daily) and myoinositol to support blood sugar balance. If appropriate, intermittent fasting can be beneficial, and consistency is key—try to eat at the same times each day. Having your last meal of food intake earlier in the evening may also help with insulin regulation. To stay on top of your health, work closely with your doctor to monitor blood sugar levels and ensure early intervention if necessary, particularly to prevent the development of pre-diabetes or diabetes.

Checklist for Your Hormonal Health

- **Serum Hormone Testing:** You might consider testing your blood for estradiol (E2), estrone (E1), progesterone, testosterone, DHEA-S, cortisol, insulin, leptin, and thyroid hormones (free T3, free T4, TSH). Testing sex hormone-binding globulin (SHBG) can also be helpful.

- **Track Symptoms:** It is useful to keep track of symptoms like sleep issues, mood swings, energy levels, cognitive changes, and menstrual cycle irregularities. A hormone tracker app or symptom journal can help.

- **Urine Hormone Metabolite Testing:** A 24-hour urine hormone metabolite test could provide insight into how your body processes hormones like estrogen, progesterone, and cortisol.

- **Maintain a Healthy Gut Microbiome:** Probiotics or a fiber-rich diet may support a healthy gut.

- **Diet for Hormone Detoxification:** Including cruciferous vegetables (like broccoli or kale) in your diet could help support estrogen detoxification.

- **Bone Health and Cardiovascular Risk:** A DEXA scan can help monitor bone density, especially after menopause. Assessing your cardiovascular risk with lipid panels or coronary CT scans could also be important. Heart disease is the number one killer in women as well as men.

- **Bioidentical Hormone Therapy (BHRT):** If you're exploring BHRT, using bioidentical estradiol and progesterone in cyclical patterns might be helpful. Regular testing can help fine-tune the dosage.
- **Nutrient Support:** You might consider supplements like magnesium, Vitamin D3, omega-3 fatty acids, and B vitamins to support hormone balance and overall health.
- **Regular Follow-Ups:** Schedule follow-ups every few months to recheck hormone levels and adjust your plan as needed.
- **Monitor Insulin Levels:** This is especially important if you have had gestational diabetes. Insulin levels rise before blood sugar increases, so regular lab checks are crucial.
- **Maintain a Stable Blood Sugar Level**: Avoid sugar, artificial sweeteners, processed foods, and alcohol to prevent insulin spikes.
- **Adopt a Circadian-Based Eating Schedule**: Insulin production and sensitivity decrease later in the day. Eat your meals earlier, with lunch as the main meal, to align with this natural rhythm.
- **Post-Meal Activity**: Walk for at least 10 minutes after each meal to help maintain insulin balance. This will drive the sugar from your bloodstream into your muscles.
- **Check Leptin Levels**: Blood tests can help determine your leptin status and provide insight into your hormonal health.
- **Monitor for Leptin Resistance**: If you feel persistent hunger or tend to overeat, you might be experiencing leptin resistance. This could increase the risk of weight gain and related issues.
- **Diet and Lifestyle Adjustments**: Prioritize balanced meals and avoid factors that may contribute to leptin resistance.
- **Maintain Metabolic Health:** Follow longevity habits like balanced nutrition, exercise, and stress management to support reproductive health and possibly delay menopause.
- **Avoid Environmental Risks**: Minimize exposure to smoking, poor nutrition, stress, and weight extremes, which accelerate ovarian decline and impair fertility.

Reproduction

> "Humans are one of only five species on Earth that undergo menopause, including killer whales, short-finned pilot whales, beluga whales, and narwhals."

Accelerated Ovarian Aging Compared to the Rest of the Body

Ovarian aging refers to the progressive decline in a woman's ovarian reserve—the number of oocytes or eggs she has left. Unlike other cells in the body, women are born with a finite supply of eggs. Although a small number are released during menstruation, most undergo atresia, a process where the eggs die and are reabsorbed by the body. This decline in ovarian function occurs approximately five times faster than in other organ systems, a phenomenon known as accelerated ovarian aging. Incredibly, at about 26 weeks of gestation (while in utero), a female fetus possesses roughly 6 million eggs (the most she will ever carry), a number that drops to about 1 million by birth and dwindles to 400,000 when she enters puberty and her reproductive years.

Eggs are housed within ovarian follicles, which contain cells responsible for estrogen production. By around age 54, most women have depleted their ovarian reserve and enter menopause (defined as going a full 12 months without menstruation) despite having released fewer than 500 eggs in their lifetime. While ovarian function declines, other body systems—like the heart, brain, muscles, and lungs—continue to function relatively well, though not without some age-related reduction in efficiency. For example, brain function may exhibit only minor cognitive decline, and muscle mass and lung capacity typically remain around 80 percent of their peak levels.

With the onset of menopause, a sharp decline in estrogen levels triggers a variety of symptoms, including mood swings, hot flashes, vaginal dryness, reduced libido, sleep disturbances, and weight gain. At this stage, natural conception is no longer possible. Ironically, menopause also accelerates aging in other systems of the body. Lower estrogen levels alter metabolism, leading to an increased deposition of fat in blood vessels and a higher risk of cardiovascular disease.[121] Bone density also decreases, resulting in osteoporosis, a condition where bones become fragile and more prone to fractures.

Although longevity practices aimed at maintaining metabolic health can help support fertility and alleviate some menopausal symptoms, they cannot entirely prevent ovarian aging.

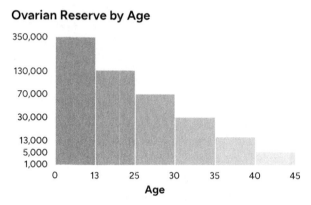

Source: W. Hamish B. Wallace human ovarian reserve from conception to menopause.

In addition to a reduction in egg quantity, the quality of the remaining eggs declines with age, further impairing fertility. Ovarian decline is influenced by the initial number of eggs and the rate at which they degrade. As with many biological processes, genetics play a significant role in determining ovarian longevity, though environmental factors can also have a profound impact. Smoking, poor diet, chronic stress, inadequate sleep, and being either underweight or overweight are all associated with a faster reduction in ovarian reserve and poorer fertility outcomes.

Healthy fertility relies on viable eggs and a body capable of supporting pregnancy. Without these, the risk of complications during pregnancy increases. Poor egg quality is linked to a higher likelihood of miscarriage and genetic abnormalities in offspring. Fertility challenges tend to become more common after age 30, with pregnancy complications rising as well. Practicing healthy longevity habits remains the most effective way to mitigate environmental impacts on ovarian decline, safeguard fertility, and delay the onset of menopause.

Egg Freezing

While healthy longevity practices can greatly benefit female fertility, they are often not enough to counter the rapid pace of ovarian decline, especially for women who wish to start a family in their early 30s or beyond. Additionally, some women face premature ovarian insufficiency, a condition marked by an earlier-than-expected decline in ovarian function and the onset of menopause. Since a woman's eggs are typically abundant and healthy in her 20s, egg freezing is one of the most effective ways to preserve fertility. This allows a woman to have a healthy pregnancy later in life, using eggs frozen at a younger age (e.g., 25 years old).

Despite being a reasonably effective option, egg freezing does have its challenges and remains impractical for widespread use, even though it's been practiced since the 1980s. The main limitation lies in the fact that only mature eggs can be frozen, and obtaining mature eggs has historically been difficult. In a woman's body, eggs are stored in an immature state within the ovaries and only mature in preparation for ovulation during menstruation. Since mature eggs are the only ones that can survive the freeze-thaw process and be fertilized with sperm, the procedure requires eggs to be matured beforehand.

To mature eggs for extraction, women undergo a two-week regimen involving approximately 24 hormone injections designed to stimulate egg development. This hormonal treatment prompts the ovaries to release mature eggs, which a fertility specialist can then retrieve through an ultrasound-guided procedure.

However, this process is costly and time-consuming, requiring significant commitment from the patient and fertility clinic. Additionally, the hormone treatments used to stimulate egg maturation can lead to ovarian hyperstimulation syndrome (OHSS), a condition that affects about 33 percent of women undergoing the procedure. Symptoms of OHSS include abdominal bloating, pain, and nausea, and in more severe cases—around 6 percent of women—these symptoms can be pronounced. About 1–2 percent of women may even require urgent hospital care. The long-term health effects of these treatments remain unclear.

Women with large ovarian reserves, such as those opting to freeze eggs for voluntary fertility preservation, are at higher risk of developing OHSS, as they have more follicles that respond to the hormone treatments. These challenges, including the risks and side effects, contribute to egg freezing not yet being a common practice for fertility preservation.

It's also worth mentioning the cost of this process, which can be prohibitive. The basic egg-freezing cycle can run between $7,000 to $12,000 and includes an initial consultation, hormone stimulation medications, monitoring appointments, and the egg retrieval procedure. You may also need to pay for medications ranging in price from $2,000 to $5,000. Add to this an annual storage fee of $500 to $1,000 per year. There are other potential hidden costs such as pre-testing blood work, anesthesia, and potentially the genetic testing of retrieved eggs, bringing the total cost of egg freezing to somewhere between $10,000 and $20,000 for a single cycle, plus the future fertilization and implantation of those eggs.

Gameto's Recent Advances in Fertility Treatments

Reducing the amount of hormones administered is crucial to advancing fertility treatments. One alternative approach is in-vitro maturation or IVM, a technique that involves a much smaller hormone regime—typically just three injections over three days—to extract immature eggs.

These eggs are then matured outside the body in a laboratory dish. Although IVM has been practiced since the 2000s, its pregnancy success rates have not been high. This is largely due to the static liquid environment used to mature the eggs, which cannot replicate the dynamic conditions of a living ovary. In the body, eggs mature within ovarian tissue, which can respond to their needs and continuously provide the necessary nutrients—something static liquid solutions fail to do.

Dina Radenkovic, MD, Co-Founder & CEO, Gameto

Gameto, a biotech company led by Dr. Dina Radenkovic (and a company that Peter advises and has invested in), is pioneering solutions to improve women's reproductive health at every stage of life. Their innovations span egg freezing, infertility treatments, diseases of the reproductive years, and menopause, with a broader focus on addressing the conditions that arise after menopause. Gameto is revolutionizing traditional fertility processes through advanced therapeutics.

One of Gameto's breakthrough products is Fertilo, an IVM solution made from engineered ovarian support cells. Fertilo effectively mimics the natural ovarian environment outside the body, creating an "ovary in a dish" scenario. When immature eggs are placed in this maturation solution, the success rate for egg maturation improves by roughly 30 percent compared to using static liquid solutions alone.[122] This innovation has the potential to significantly enhance IVM's effectiveness, allowing women to undergo egg freezing and

IVF with fewer hormone treatments. This would reduce both the cost and side effects of fertility treatments while improving safety.

The offerings developed by Gameto have wider applications beyond fertility treatment. One such application is Ameno, an implant designed to produce estrogen for women who can no longer produce it naturally, offering relief from menopausal symptoms without the need for traditional hormone replacement therapy.

Another application, Deovo, serves as a platform for drug developers to test the effects of new medications on the ovaries. This technology could play a key role in bringing more medications to the market that are safe for young women, eliminating uncertainties around the long-term effects of certain medications on fertility.

The growing emphasis on women's health—particularly reproductive health—is reversing decades of inattention. These advancements are starting to yield positive results, contributing to a brighter future for women's health care.

Nourishment

Viewing food as nourishment rather than just a source of calories can help many women redefine and heal their relationship with eating. It's important to recognize that everyone's nutritional needs are different, and what works for one person may not work for another. To better understand your unique requirements, consider exploring food sensitivity and gut microbiome testing. Emotional and psychological support can also be key in fostering a positive relationship with food. As Marc David, the founder of the Institute for the Psychology of Eating, says, "Our relationship with food mirrors our relationship with life." Some live to eat, and others eat to live. Ultimately, nutrition goes beyond providing energy; it sends important signals to your cells that influence overall health and well-being in addition to its social and pleasure aspects.

A well-rounded, nutrient-dense diet is fundamental to good health. Incorporating plenty of color-rich vegetables, lean protein, and fiber-rich foods can be beneficial, with strong evidence supporting the Mediterranean diet. Ultimately, a diet based on whole foods is best for almost everyone. However, nutrition can be a personal journey; experimenting with what makes you feel good, especially in alignment with your genetic or ancestral background, is a great way to find what works best for your body. Vegetables like broccoli, cabbage, and cauliflower are particularly helpful for supporting your body's detoxification processes, while spices such as turmeric, rosemary, and oregano can provide additional anti-inflammatory and antioxidant benefits.

In terms of daily eating habits, aiming for three balanced meals a day helps maintain consistent energy levels, which supports regular exercise and overall activity. While fasting can be useful for certain health goals, ensuring you're still getting the nutrition your body needs is important. Green tea is another great addition to your routine, offering a boost of antioxidants. Additionally, sipping on organic, grass-fed bone broth can help heal the gut and promote collagen production, supporting digestive and skin health.

Taking the time to truly savor your food can also make a difference. Beautifully presenting your meals and eating mindfully engages your parasympathetic nervous system, which is crucial for optimal digestion. Exploring new foods, including fermented options, is another way to boost health; fermented foods aid in the production of short-chain fatty acids that support the gut lining and gut-brain connection and strengthen the immune system. Lastly, when you're feeling tired, try reaching for a calming cup of Tulsi tea instead of coffee or sweets. Tulsi helps balance adrenal function and provides a more gentle, sustained energy boost.

To close this section, it's worth mentioning that I encourage food sensitivity testing, gut microbiome testing, and, if appropriate, emotional and psychological support to develop a healthy relationship with food. Nutrition is not just about calories; it is a signal to your brain and your cells supporting overall physical and psychological health.

Fasting—Is it right for you?

Intermittent Fasting and Women: A Different Approach

Intermittent fasting has gained significant popularity in recent years, with endorsements from celebrities and healthcare professionals. Research points to benefits such as weight loss, improved blood sugar control, and reduced inflammation. However, an important question arises: Is intermittent fasting equally beneficial for women?

Before embracing intermittent fasting or any longevity practice, it's essential to consider whether it is appropriate for your unique needs as a woman. Specifically, were the studies supporting these practices conducted on women? The answer to this question can provide critical context.

A fundamental difference between men and women lies in fasting insulin levels. Women generally have lower fasting insulin, which means we experience blood sugar regulation differently from men. This difference is just the beginning of how intermittent fasting can uniquely affect women, particularly when it comes to hormonal balance.

Let's explore the hormonal shifts that intermittent fasting can trigger in women.

Prolonged fasting may impact estrogen and progesterone, the two key sex hormones. Extended fasting can disrupt the delicate hormone cycles that govern the female body, potentially leading to menstrual irregularities. In some cases, women have reported disruptions in their menstrual cycles, with some experiencing cessation altogether. But why does this happen? One theory suggests that the body perceives fasting as a signal of food scarcity. During famine, the body may suppress ovulation to prevent pregnancy—a survival mechanism designed to avoid reproduction when food is scarce. This evolutionary response, while protective, can cause significant disruptions in hormonal balance.

In practice, lowered levels of estrogen and progesterone due to intermittent fasting can result in a variety of symptoms, including:

- Irregular or missed periods
- Mood swings
- Night sweats and hot flashes
- Decreased libido
- Sleep disturbances
- Dry skin and hair loss
- Heart palpitations
- Infertility

However, the impact of intermittent fasting can change post-menopause. Once women reach menopause, typically around age 50, estrogen and progesterone levels remain low and relatively stable. *For postmenopausal women, intermittent fasting may actually be more effective.* One study even showed that intermittent fasting could help with weight loss in a population of postmenopausal women.[123] Still, caution is necessary, as this result was not consistently observed across all studies.

The important takeaway is that intermittent fasting, like any longevity practice, should be approached with careful consideration of a woman's unique physiology. For example, if you're pregnant, trying to conceive, or experiencing fertility issues, intermittent fasting should be avoided. The same applies if you are underweight or have a history of eating disorders. Additionally, fasting can lead to muscle loss, as the body may use muscle as an energy source. Given that women generally have lower muscle mass than men, this can pose an even greater concern. Extended fasting may also increase cortisol levels, triggering a stress response that accelerates the breakdown of muscle, bone, and connective tissue.[124]

I aim for time-restricted eating, stopping food intake two hours before bed, fasting for 12 to 14 hours between the last meal (dinner) and breaking my fast the next day. When I'm stressed or engaging in more intense workouts, I shorten the fasting window to about 10 to 12 hours to prevent overburdening my body.

Toxins

The Unseen Impact of Toxins on Women: My Approach

In our ever-evolving world, awareness of environmental toxins and their impact on our health continues to grow. As women, we are particularly vulnerable to these toxins, and there's a clear reason for this: Many harmful substances accumulate in fat tissue, and since women naturally have higher body fat percentages than men, we tend to store more of these toxins.

Daily routines often exacerbate this issue. On an average day, *a woman is exposed to as many as 168 chemicals* through beauty and personal care products. From nail polish and skin creams to shampoos and perfumes, the list of potential chemical exposures is extensive. Societal pressures to adhere to certain beauty standards further amplify this exposure, as we inadvertently absorb more chemicals in pursuit of these ideals.

Many of these toxins contain xenoestrogens—external agents that mimic the function of natural estrogen in the body. The body does not synthesize these chemicals, yet they are found in everyday items such as plastics, tap water, canned food, and even laundry products. Xenoestrogens bind to estrogen receptors, producing effects similar to those of natural estrogen, often leading to a condition known as "estrogen dominance." Symptoms such as breast swelling and pain may not solely be a result of the body's natural hormones but are often influenced by these external disruptors.

To reduce my exposure to these invisible threats, I've taken a proactive approach. I choose personal care products that are organic and free from harmful additives like parabens, phthalates, and acrylamides—although there are many other toxic compounds to avoid. For a more comprehensive list of harmful chemicals, visit SafeCosmetics.org/chemicals. Additionally, I prioritize organic food to limit my exposure to dietary toxins such as herbicides and pesticides, which can also disrupt hormone function.

Our understanding of the health impacts of environmental toxins continues to expand. In medical school, we were taught that men rarely experienced bone density loss before old age. Yet today, bone density loss is being observed in men as young as 30, and it remains an even more prominent issue for women. While the exact causes are still being researched, there is growing suspicion that toxins play a significant role.

The truth is that toxins are pervasive in our world. However, my approach is simple: *Focus on what you can control and let go of stress over what you can't.* While it's impossible to eliminate every toxin, being informed and making conscious choices can significantly improve your health and well-being.

Detox Practices Checklist

- **Induce Sweating for Toxin Elimination**: Regular physical exercise and using infrared or traditional saunas can facilitate detoxification through sweat, promoting the excretion of heavy metals, PCBs, and other toxins.

- **Select Non-Toxic Cosmetics and Household Products**: Prioritize personal care and cleaning products free from endocrine-disrupting chemicals (EDCs) like parabens, phthalates, and triclosan. Use databases such as the Environmental Working Group's (EWG) Skin Deep and EWG's Guide to Healthy Cleaning for safer alternatives.

- **Incorporate Dietary Fibers and Cruciferous Vegetables**: Consuming a diet high in soluble fiber (e.g., from oats and legumes) aids in the binding and eliminating toxins via bile excretion in the gut. Cruciferous vegetables (broccoli, kale, cauliflower) enhance phase II liver detoxification by upregulating the production of glutathione and other detoxifying enzymes.

- **Supplement with Detoxification-Enhancing Compounds**: Take N-acetylcysteine (NAC), which acts as a precursor to glutathione—one of the body's most potent antioxidants involved in detox pathways. Additionally, citrus pectin can bind heavy metals and toxins in the gastrointestinal tract, supporting their safe excretion.

- **Professional Detox Protocols**: For individuals exposed to high levels of heavy metals (e.g., lead and mercury) or environmental toxins (e.g., mold and VOCs), work with a functional medicine doctor or toxicologist to initiate a medically supervised detoxification protocol. This may include chelation therapy, liver support, and lymphatic drainage treatments.

Stress

Dealing With Stress and Anxiety as a Woman

It's widely recognized that anxiety and stress are more prevalent among women than men. The sheer number of responsibilities women often juggle can feel overwhelming, and the resulting stress affects every part of us—from our mental well-being to our very cells. Self-care is not a luxury; it's a necessity for maintaining resilience and well-being. However, the effects of stress go beyond the immediate feeling of unease. Chronic stress is often accompanied by exhaustion, anxiety, panic, and anger. If left unchecked, these symptoms can lead to burnout and, ultimately, a breakdown.

Stress also has a profound impact on our endocrine system, especially our hormones. One of the important ways that this happens is through what's known as progesterone or cortisol steal, depending on how you look at it.

Women in the Okinawa Blue Zone form tight-knit friendship groups and have 40% higher chance of reaching 100 years old than Japanese men.

Under stress, the body prioritizes the production of cortisol—the stress hormone—over sex hormones like progesterone, which can disrupt the delicate hormonal balance women need for overall health. Since progesterone and cortisol are made from the same precursors, the need for more cortisol often results in less available progesterone.

That said, stress isn't always the enemy. In manageable doses, stress can be a powerful motivator. It pushes us to perform, providing the drive we need to thrive. However, finding that "sweet spot" is crucial—too little stress and we lack energy and purpose; too much and we risk physical and emotional derailment.

Having faced stress throughout my life, I've worked hard to not internalize external pressures. While I don't always succeed, making a conscious effort to avoid taking things personally has been invaluable. I've also found that binaural beats (search for a soundtrack on YouTube to try it out)—a technique that uses slightly different auditory signals in each ear to synchronize brain activity—can help entrain my brainwaves into calming patterns, such as alpha, delta, or theta waves.

Additionally, nature is one of my greatest stress relievers. Walking outdoors provides a sense of calm and peace. Just as important, though, is maintaining and nurturing key relationships. Social support is critical for managing stress and anxiety, so be sure to cultivate connections with those around you.

It's also important to learn to recognize what stress feels like in your body, gut, or chest and to train yourself to take a few moments when you feel those symptoms to take a deep breath, which activates your parasympathetic system. Try it now; *just a couple of deep breaths can be transformative.*

While it's important to optimize health and longevity through practices like these, it's also crucial to accept that we won't always be perfect. Sometimes, even when we know something is beneficial, it may not feel enjoyable. For instance, I know cold plunges can positively affect longevity, but I don't do them often because I find the cold incredibly stressful. Even thinking about an ice-cold bath makes my body tense.

Vagal Tone

As discussed in an earlier chapter, stimulating the vagus nerve is a powerful way to enhance the body's parasympathetic response, helping balance stress. For women, spending time with friends is a great way to boost vagal tone. Building community and socializing can provide emotional support and improve overall health. Staying connected with others is key to reducing stress and boosting well-being.

Trauma

Difficult childhood or early adult experiences can have a lasting impact on our adult lives. While the full discussion of trauma is beyond the scope of this book, it's important to recognize the importance of this unseen and often unspoken cause of poor health. If this is true for you, *please take the time to explore this, at a minimum, with a close friend and, ideally, with a professional.* Please don't put it off any longer. Seeking professional help to address these deeper emotional wounds is always a good idea.

Checklist for Navigating Your Stress

- **Monitor Heart Rate Variability (HRV)**: Use HRV as an indicator of how well your body shifts from sympathetic (fight or flight) to parasympathetic (rest and digest) states. You can find this on an Apple Watch (health kit) or an Oura Ring.
- **Increase Vagal Tone**: Practice box breathing, sing, or spend time with friends to stimulate the vagus nerve and enhance parasympathetic activation.
- **Recognize Stress and Breath**: Learning to recognize the symptoms of stress in your body and, thereafter, taking just one minute to take some deep breaths can be transformative.
- **Meditate**: Regular meditation can help regulate stress by promoting mindfulness and reducing cortisol levels.
- **Maintain Daily Rhythms**: Align your activities with your body's natural circadian rhythms to enhance balance and energy throughout the day.
- **Rest When Tired**: Listen to your body and rest when you feel fatigued to prevent overstimulation of the sympathetic nervous system.
- **Incorporate Breathwork**: Techniques like diaphragmatic breathing can quickly calm the nervous system and reduce stress.
- **Consider Acupuncture**: Acupuncture can help restore balance in the body's energy flow, reducing stress and anxiety.
- **Try Tai Chi, Yoga, or Drumming**: These physical practices help regulate the nervous system, promote relaxation, and boost mental clarity.
- **Supplement with Adaptogens**: Use ashwagandha to balance your stress response and Rhodiola to support energy levels when stress leaves you feeling depleted.
- **Sauna**: Consider adding a 15–20 minute sauna three times per week into your practice to eliminate toxins and increase your parasympathetic tone.

Exercise

Should Women Exercise Differently Than Men?

While it's universally accepted that both men and women experience significant health benefits from regular exercise, there has long been an unspoken divide in the types of activities each gender tends to pursue. Women are often encouraged to focus on cardio, yoga, and fitness classes, whereas men are directed toward weightlifting, competitive sports, and high-intensity training. However, as we begin to better understand the role of muscle mass in health and longevity for both genders, a notable disparity remains: only 17 percent of women regularly lift weights, while they outnumber men in yoga and group fitness classes by a ratio of 5:1.[125]

This raises an important question: Does the difference in exercise preferences between men and women stem from physiological factors, or does societal conditioning drive it?

While gender stereotypes certainly play a role, there are both similarities and differences in how men and women respond to exercise. Men tend to have greater overall muscle mass and larger muscle fibers compared to women, but when adjusted for body weight, women can achieve similar strength ratios with proper training. Women also tend to develop lower body muscle mass more quickly, whereas men build upper body muscle more easily.

Women may benefit from shorter rest periods between sets during weight training, as they often recover faster. Regarding metabolism, women typically burn more sugar at rest but are more efficient at burning fat during exercise, particularly during weight training and high-intensity intervals. Post-exercise, women may not need as many carbohydrates for recovery and may fare better with a balanced meal, as opposed to the higher carbohydrate refueling often recommended for men. Additionally, while men tend to respond well to calorie deficits and high-intensity training for fat loss, women are more likely to trigger a stress response, leading to cortisol release and, potentially, fat storage with intense training or severe calorie restriction.

For my fitness routine, I prioritize consistency and balance. My base exercise is Zone 2 training (explained by Peter in the chapter on exercise), which involves fast walking outdoors, allowing me to enjoy nature and catch up on podcasts or audiobooks. I incorporate heavy weight training and dynamic heavy kettlebell training two to three times per week, often aiming for short, intense bursts of activity lasting 1–2 minutes to challenge my VO2 max (also in the chapter on exercise). Recovery and stretching are equally important, as is my go-to post-workout Ka'chava shake.

Despite yoga's association with traditional female fitness routines, I make a point to practice it daily for flexibility and stability. To add more movement to my day, I use a standing desk and regularly practice tree pose during meetings. I also make it a habit to do 20 squats after every meeting to boost circulation and stabilize blood sugar. Additionally, I've found that static muscle contractions during meetings are a discreet but effective way to build strength.

Pelvic Floor Health

One often overlooked area for women is pelvic floor health. Many women experience weakened pelvic floor muscles, which can make high-intensity exercise uncomfortable or difficult. If this is a concern for you, seeing a gynecologist or pelvic floor specialist is essential. Addressing pelvic floor weakness not only improves your participation in sports but can also enhance your sex life.

Exercise Routine Checklist

- **Incorporate High-Intensity Interval Training (HIIT)**: HIIT exercise boosts cardiovascular fitness and fat-burning efficiency and stimulates the formation of more healthy mitochondria.
- **Strength Train Regularly**: Focus on weightlifting to build and maintain muscle mass, which is critical for metabolic health and longevity.
- **Prioritize Functional Mobility and Movement**: Explore resources like *The Ready State* to improve movement patterns and prevent injuries.
- **Incorporate Yoga**: Yoga helps with flexibility, stability, and stress management.
- **Pilates for Bone Health**: Pilates can enhance bone density and improve posture, which is especially important for women.
- **Try Swimming**: Swimming provides low-impact conditioning and can be a form of moving meditation.

Sleep

Do Women Really Need More Sleep Than Men?

The importance of sleep for longevity has been well established, but is there a difference in how much sleep women need compared to men? Research suggests that, on average, women require about 20 more minutes of sleep per night than men.[126] Additionally, women tend to get more deep sleep. Circadian rhythms also differ between the sexes; women's circadian clocks often shift earlier, meaning you may feel sleepier in the evening and wake up earlier in the morning. Hormonal fluctuations also play a role, with many women experiencing better sleep quality in the first half of their menstrual cycle when estrogen levels are higher. Hormones can also influence body temperature, which in turn affects sleep. Using a cooling mattress that adjusts to your monthly cycle can be an excellent way to manage these temperature changes and improve sleep quality.

Peter's chapter on sleep applies equally to women, and I encourage you to read it. The only comment I might make is that many women usually don't enjoy a room temperature below 65 degrees Fahrenheit and prefer a hot pre-bed bath to help lower their core temperature in the hours after bathing and during the initial stages of sleep.

Checklist for Better Sleep

- **Red Light Therapy**: Promotes relaxation and improves sleep quality.
- **Epsom Salt Baths**: Aids in muscle relaxation and helps calm the nervous system.
- **Magnesium Glycinate**: Supports cognitive function, detox, and promotes deeper sleep.
- **Phosphatidylserine**: Helpful if stress is disrupting your sleep by lowering cortisol levels.
- **Lemon Balm/Valerian Root Tea**: Natural remedies that help induce relaxation and improve sleep onset.
- **Cultivate a Nighttime Routine**: Consistency in your bedtime routine can help regulate your circadian rhythm. (See Peter's chapter on routines.)
- **Wear Blue Light-Blocking Glasses**: Reduce exposure to blue light, helping maintain a healthy circadian rhythm.

Skin and Hair Health

Once again, hormone balance plays a key role in maintaining vibrant skin and healthy hair. Sex hormones like estrogen and testosterone significantly influence the structure and quality of skin and hair. For women, ensuring adequate intake of micronutrients, particularly iron, is crucial for hair health, as is maintaining good thyroid function. Incorporating sea vegetables into your diet can enhance hair shine, thanks to their mineral content. "Sea vegetables" is a term used to refer to various edible seaweeds and algae that grow in saltwater, including nori (used in sushi rolls), kelp, dulse, wakame, kombu, spirulina, and chlorella. Sea vegetables are known for their high mineral content, particularly iodine, iron, calcium, and magnesium. They also contain various vitamins and antioxidants.

Another important reminder is that balanced estrogen levels and healthy testosterone contribute to hair follicle stimulation and hair growth.

Sleep also has a significant impact on skin health. While we sleep, the glymphatic system works to clear toxins from the brain, functioning like a pump that helps remove waste, supported by the rhythmic action of the heartbeat. This process is essential for the release of growth hormones overnight, which aid in skin repair and regeneration. Maintaining a regular sleep rhythm, stable blood pressure, and good vascular health is crucial to facilitate this nightly detoxification and promote skin health.

Regarding skin care products, peptides like GHK-Cu can be used topically or in capsule form to promote skin regeneration. Exosome-based products are effective in repairing oxidative damage to the skin. Additionally, using NAD topically or undergoing PRP (platelet-rich plasma) facials can also help repair skin tissue.

To round out my top-level recommendations related to skin and hair, here are four easy-to-utilize but often ignored strategies:

Consume Collagen: Whether you obtain your collagen from food or supplements, it is a key to maintaining skin elasticity and hydration.

Sunscreen: Proper sun protection, especially during peak hours, helps preserve skin health and prevent premature aging. Have it easily within reach and use it, and if you don't already, take a moment, go on Amazon, and order a high-quality SPF 40 product.

Drink Water: Hydration is equally important; drinking water with minerals, such as humic and fulvic acids, enhances cellular hydration.

Red Light Therapy: Red light therapy is another excellent method for improving skin health. According to research, the consistent use of red light

Bone Health

Bone Health and Menopause

Bone is a dynamic tissue, constantly undergoing remodeling, much like other tissues in the body. This continuous process is tightly regulated by hormones, inflammatory cytokines, and various cell types involved in bone metabolism. Hormones and mechanical stress play significant roles in maintaining bone health. During adolescence, we lay down the foundation for bone strength that will serve us throughout life, and questions remain about how the use of birth control might impact bone quality during this critical developmental phase. As women age, bone continues to remodel, highlighting the importance of lifelong exercise, balanced hormones, and proper mineral nutrition.

As we approach menopause, the effects on our bone health become even more pronounced. The Study of Women's Health Across the Nation, known as the SWAN Study, identified menopause as a critical period for changes in bone strength, setting the stage for osteoporosis and increased fracture risk in older age.[127] This study suggests that the menopausal transition represents a time-sensitive window of opportunity to intervene, preventing rapid bone loss and minimizing the microarchitectural damage that can lead to osteoporosis in later years.

Here are key points from the paper:

- Bone resorption begins increasing two years before the final menstrual period (FMP), peaks approximately 1.5 years after the FMP, and then stabilizes.
- The rapid phase of bone loss occurs over three years around the FMP, coinciding with increased bone resorption.
- Metabolic factors during menopause—such as insulin resistance, inflammation, and obesity—are linked to reduced bone strength and heightened fracture risk.

Given these findings, it is critical to work closely with a physician to monitor your hormone levels and bone health throughout this transition.

Regular testing, such as a DEXA scan, can detect early changes in bone density, helping catch issues before they progress. Hormone replacement therapy may also be considered to support both overall health and bone strength. Additionally, challenging your bones through weight-bearing exercises, reducing inflammation, and ensuring adequate micronutrient intake all contribute to a comprehensive bone health strategy.

Checklist for Strong Bones:

- **Lift Weights**: Engaging in weight-bearing exercises helps maintain and improve bone density.
- **Vibration Plate Training**: Using a vibration plate can enhance bone strength through mechanical stimulation.
- **OsteoStrong**: Explore programs like OsteoStrong, which focus on resistance-based exercises to improve bone health.
- **Ensure Adequate Mineral and Vitamin Intake**: Essential minerals like calcium, magnesium, and Vitamin D are crucial for bone formation and maintenance.
- **Creatine**: Consider adding creatine to your supplements. It offers broad benefits for women, supporting muscle and brain health and bone strength.
- **Optimize Gut Health and Toxin Clearance**: A healthy gut ensures proper nutrient absorption while detoxifying pathways, reducing inflammation, and supporting overall bone health.
- **Consult a Healthcare Provider**: If your bones require additional support, work with your doctor to explore advanced interventions or therapies.

Supplements and Medications

What Do I Take? It Depends.

When people ask me what supplements I take, I always say, "It depends." I don't take the same supplements every day. I typically buy a range of supplements, keep them at home, and choose the supplements to take for a given week based on how I feel.

I recently took an inventory of all the supplements I have in my cupboard, and I've included a summary of them below. **Note:** I've tried to group these into categories, but many of these supplements will overlap across categories. I also try to use products that contain many ingredients in one capsule to limit the number of capsules I take.

Remember that the dosing of supplements should be modified for women and is unique to you, your physiology, and desired outcomes. Based on our smaller body size and unique metabolism, the standard dosing determined for men may not be applicable for us, and we often need lower doses. Ultimately, dosing and frequency will be determined between you and your physician. Consider using this list as a reference for discussion with your medical doctor. More details on many of these supplements can be found in Peter's chapter on supplements.

Gut and Microbiome Supplement Support:

- **Digestive Enzymes with Food** (typically with each large meal): Our ability to make stomach acid decreases as we age. Stress also decreases our ability to make stomach acid, so I take a digestive enzyme to ensure I get all the benefits of the food I eat and can fully digest and absorb all the nutrients.
- **Butyrate**: Butyrate is a short-chain fatty acid made from fiber by the microbes in our gut. It keeps the lining of our GI tract healthy and prevents leaky gut. By taking butyrate, I can support my gut lining and create an environment for the good microbes in my gut to thrive.
- **TUDCA**: Tauroursodeoxycholic acid, or TUDCA, is a bile acid derivative that occurs naturally in the body and is made by microbes when they metabolize our bile acids that are released into our intestines by our liver and gallbladder. TUDCA has been shown to support brain function, liver, kidney, eye, mitochondria and cellular health, insulin sensitivity, gut microbiome balance, and bile flow, which help digestion of fats.
- **Fiber**: I take a variety of different fiber supplements. Fiber is one of the most important things for maintaining a healthy microbiome. It acts as a fuel for the beneficial microbes. Just like with our food, it is important to get a variety of different fibers, so I will rotate several supplements with different types of fiber. I usually add it to a shake or sprinkle on my food.

- **N-Acetyl-Glucosamine:** Provides building blocks for the intestinal glycocalyx, which is the inside coating of the gut that protects us from leaky gut. NAG has also been shown to increase the elasticity of tissues surrounding blood vessels and support immune health.

Cardiovascular Related Supplements:

- **Nattokinase:** Nattokinase is a very safe blood thinner that prevents clotting and can be helpful if you travel a lot or do a lot of sitting to prevent blood clot formation.
- **Arterosil:** Arterosil is a glycocalyx regenerating compound and has been shown to build and maintain the endothelial glycocalyx. This is the inner coating of all our blood vessels. When this structure breaks down, it is the first step in atherosclerosis and damage to our vessel's walls.
- **Vascanox:** Vascanox increases nitric oxide production, which is needed to dilate our blood vessels. This can maintain healthy blood vessels and blood pressure.
- **Omega-3 Fatty Acids:** Omega-3 fatty acids are a component of our cell membranes. They are precursors to the anti-inflammatory molecules of our immune system. The fatty acids we take need to be balanced. We need the right ratios of omega-3 to omega-6 and saturated fatty acids. Most of us do not get enough omega-3 fatty acids. The right balance benefits our brains and hearts and reduces overall inflammation. It is important to test your omega-3 levels to ensure you have an adequate amount and a good ratio of omega-6 to omega-3 fatty acids in your cell membranes.

Brain Health Related Supplements:

- **Prodrome Neuro and Gila Plasmalogens:** These are important components of our cell membranes, especially in our brain. As we get older, our levels of plasmalogens decrease. Low plasmalogen levels are associated with poor cognitive function and dementia.
- **Phosphatidylcholine (PC):** Phosphatidylcholine makes up most of our cell membranes. Choline is also used to make acetylcholine, which is a critical neurotransmitter in our brains and predominantly used by our vagus nerve. PC can help with mood, memory, stress, and muscle control.

- **Ketone Esters**: Ketones are amazing fuel for our brains. We make ketones when we fast or follow a ketogenic diet. We can also take them as a supplement, protecting our brains and enhancing cognitive function. I take them whenever I need a brain boost.

Cellular / Mitochondrial Supplement Support:

- **Coenzyme Q10**: CoQ10 is essential for energy production in the mitochondria. It is also a powerful antioxidant.
- **Curcumin**: Curcumin is found in turmeric and has an enormous number of beneficial effects, most notably as an anti-inflammatory substance.
- **Multivitamins**: I take a multivitamin to supply my basic vitamin and mineral needs. I use one in food form with added greens and other phytonutrients, including resveratrol.
- **Acetyl-L-Carnitine**: Carnitine is needed to shuttle fatty acids into the mitochondria; the acetyl form can enter the brain. It helps with energy production and blood sugar regulation.

Stress and Sleep Support Supplements:

- **Phyto ADR Adaptogens**: Adaptogens are a class of herbs that help the body adapt to physical, chemical, or biological stressors. They have been used in traditional Chinese and Ayurvedic medicine for centuries. When I am under stress, these can really be helpful. I always make sure they contain ashwagandha, rhodiola, and ginseng.
- **Phosphatidylserine (PS)**: PS is part of our cell membranes. It can help manage the stress response as well as improve memory and cognitive function. It is great to help sleep on those stressful days.
- **Magnesium Glycinate**: Magnesium is essential for more than 300 biochemical reactions. We tend to use up magnesium when we are under stress. If we ever get tight muscles or constipation, it is often a sign of a need for magnesium. I will adjust my dose of magnesium based on the kind of day I have had.
- **Glycine**: Glycine is an amino acid that makes up the majority of collagen. It is also a precursor to glutathione and supports detoxification pathways in the liver. Glycine also acts as an inhibitory neurotransmitter in the brain and promotes relaxation and sleep.

- **Melatonin:** Melatonin is produced by our pineal gland and regulates our sleep-wake cycle. It helps set and keep our circadian rhythms on track. It is also an amazing antioxidant and supports eye, GI, heart, and brain health. It also protects mitochondria and has anti-cancer effects.

Toxin Reducing Supplements:

- **N-Acetyl Cysteine / Glutathione:** Glutathione is our major detoxifier and antioxidant, and in our toxic world, it is critical to maintain our levels. NAC is the rate-limiting step to making glutathione, so I always make sure I get enough to support my glutathione production.

General Longevity Promoting Supplements:

- **Glucosamine:** Glucosamine sulfate has been shown to correlate with increased lifespan in large human studies. It is associated with a reduced risk of cardiovascular disease and a lower incidence of diabetes. It supports the production of mitochondria and reduces inflammation.
- **Carnosine:** Carnosine is a dipeptide found in brain and muscle tissue. It protects against glycation and can counteract age-related cellular damage.
- **Low-Dose Lithium:** Low-dose lithium has been linked to a longer lifespan in epidemiological studies and enhances the health of neurons.
- **Taurine:** Taurine is an amino acid found abundantly in the brain, retina, heart, and platelets. It supports cardiovascular health, and recent studies have shown that it can mitigate some age-related declines in organ function.
- **TruNiagin and Niacin:** TruNiagin is a form of NR that can increase NAD+ levels, and niacin also supports NAD+ levels and healthy lipid levels.
- **Quercetin:** A flavonoid found in many fruits and vegetables. It has been studied as a senolytic. It also has strong antioxidant and anti-inflammatory effects and can protect against cellular damage.

- **Spermidine**: Spermidine is a natural polyamine found in certain foods and also made by our microbiome from fermentable fibers. It is known to induce autophagy and may also support cardiovascular health, reduce inflammation, and extend lifespan in animal models.
- **Fisetin**: Fisetin is a naturally occurring flavonoid in various fruits and vegetables, such as strawberries and apples. It has been investigated for its potential to promote longevity by acting as an antioxidant and anti-inflammatory and may also be a powerful senolytic.
- **Pterostilbene**: A natural compound found in foods like blueberries and grapes. It has been studied for its potential benefits in promoting longevity due to its ability to protect cells from damage with its antioxidant and anti-inflammatory potential.
- **Alpha-Ketoglutarate**: A compound involved in the Krebs cycle, a critical metabolic pathway in the body needed to make energy. AKG supplementation may promote longevity by influencing mitochondrial function and cellular energy production.
- **Urolithin A**: A compound made by the microbiome from substances found in pomegranates, berries, and walnuts that supports mitochondrial and muscle health and decreases inflammation.
- **Creatine**: Five grams of creatine powder once daily for muscle building and brain health.

Muscle and Skin Health Supporting Supplements:

- **Collagen**: Protein to promote skin health.
- **Whey Protein**: Good, easily digestible source of protein to help maintain muscle mass.
- **HMB**: Beta-hydroxy beta-methylbutyrate is a metabolite of the amino acid leucine. It has been studied for its potential to counteract sarcopenia by promoting muscle protein synthesis and reducing muscle protein breakdown to help preserve muscle mass and strength.

So, there you have it, a chapter dedicated to women and our health. I hope these longevity practices are helpful and inspire you to dig deeper, establish healthy routines, and explore your current and future practices with your physician and medical team.

Here's to your health and longevity!

Insights from Jennifer Garrison, PhD: Your Ovaries are the "Canary in the Coal Mine for Aging"

At Peter's 2024 Longevity Platinum Trip, Jennifer Garrison, PhD, and professor at the Buck Institute, shared her unique perspective on women's health, particularly how we need a new paradigm for understanding the role of female sex organs in healthspan and longevity. "Ovaries are the canary in the coal mine for aging," she explained, emphasizing that ovarian function is about much more than reproduction. Ovaries are critical signaling organs that communicate with almost every tissue in a woman's body, including the brain, heart, bones, liver, and muscles. This complex communications network makes ovaries central to overall health, and their premature aging impacts women's health outcomes dramatically. Garrison noted that ovaries age approximately *two and a half times faster* than the rest of the tissues in a woman's body, resulting in an extended period of poor health later in life. Women may live longer than men, but they spend significantly more years in poor health due to ovarian decline, a fact that must be addressed to improve female healthspan.

Jennifer Garrison, PhD

To address these challenges, Garrison called for a shift in how we discuss ovarian health: "With regard to ovaries, we have to change the conversation from fertility to longevity because the health of ovaries affects the entire body—not just reproduction." This reframing opens the door to a broader understanding of women's health and aging, advocating for more research and funding. Garrison highlighted the vast potential for economic return as well, noting that every $350 million invested in women's health research generates a $14 billion boost to the economy.[128] By focusing on ovarian health, we have the opportunity to not only improve longevity but also unlock tremendous commercial value in this underexplored field.

Major Health Differences Between Women and Men

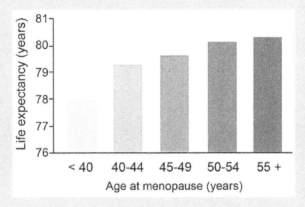

Delayed start of menopause correlates with greater female longevity. It's also true the that a brother of a woman with delayed menopause lives longer.

During Dr. Garrison's program at the Longevity Platinum Trip, we also had a chance to drill down on the major health differences between women and men, many of which are not obvious to the majority of the population. The following seven points were by far the most significant.

1. **Rate of Reproductive Decline**:[129] Men's reproductive systems decline gradually, often lasting into their 90s. In contrast, women's reproductive function rapidly falls off after the age of 30, placing women into menopause around the age of 50.

2. **Lifespan**:[130] On average, women live longer than men. The global average life expectancy is approximately 79 years for women and 72 years for men.

3. **Healthspan**:[131] While women live longer, they spend more years in poor health. Women experience 25 percent more time in poor health compared to men, largely due to chronic diseases.

4. **Disease Prevalence**:[132] Women represent 78 percent of all autoimmune disease patients and 66 percent of Alzheimer's disease cases. Additionally, women are 50 percent more likely to die within a year of having a heart attack compared to men.

5. **Research Bias:**[133] Historically, the male body has been used as the biological standard in medical research. Before 2016, 80 percent of NIH-funded studies included *only male participants*, ignoring sex differences in drug responses and disease outcomes.

6. **Drug Safety:**[134] Eighty percent of drugs pulled from the market by the FDA due to safety concerns were removed because of adverse effects experienced by women.

7. **NIH Funding:**[135] In 2023, the NIH budget was $47.68 billion, but only $4.6 billion (less than 10 percent) was dedicated to all women's health. And a mere $0.1 billion was allocated to female reproductive aging.

Appendices

Appendix A
Blue Zone Wisdom

APPENDIX A

The Blue Zones are regions of the world where people are known to live longer, healthier lives compared to the rest of the population. These areas have been identified through research conducted by the Blue Zones Project. Based on research by Dan Buettner, a National Geographic Fellow and *New York Times* best-selling author, five cultures of the world—or blue zones —were identified with the highest concentration of people living 100 years or older.

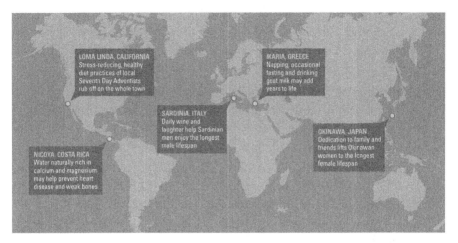

Map showing 5 blue zones around the world where people live the longest. (Webmd.com)

18 Blue Zone Secrets for a Longer Life:

The following are 18 identified hallmark practices of centenarians living in Blue Zones. How many are you following?

1. **Protect Your DNA**: Safeguard your genetic material by avoiding exposure to harmful toxins and adopting a healthy lifestyle. Use sunscreen and avoid too much sun exposure. Preventing DNA damage promotes overall longevity.
2. **Play to Win**: Approach life with a positive and competitive mindset, setting goals and challenges for yourself. This is important because it keeps your mind engaged, fosters resilience, and adds purpose to your existence.
3. **Make Friends/Community**: Cultivate meaningful relationships and engage with your community regularly. This fosters a support system that provides emotional well-being and a sense of belonging, which are essential for a long, fulfilling life.

4. **Choose Friends Wisely**: Surround yourself with people who uplift and support you in your journey. Your social circle profoundly influences your habits, attitudes, and outlook, impacting your overall health and longevity.

5. **Quit Smoking**: Break free from the grip of tobacco to protect your respiratory health and reduce the risk of life-threatening diseases. If you smoke, quit now.

6. **Embrace the Art of the Nap**: Incorporate short, restorative naps into your routine to recharge your body and mind. Napping enhances productivity and mental clarity.

7. **Follow a Mediterranean Diet**: Prioritize a diet rich in vegetables, whole grains, healthy fats, and fish. The Mediterranean diet is associated with lower rates of chronic diseases and promotes heart health, increasing your chances of longevity.

8. **Eat Like an Okinawan**: Model your eating habits after the Okinawan diet, emphasizing plant-based foods and lean protein sources. This dietary choice is linked to exceptional longevity and vitality.

9. **Get Hitched**: Cultivate a loving, committed relationship or marriage. Married individuals, on average, live longer, with a mortality rate that is about 15 percent lower than unmarried individuals.

10. **Lose Weight**: Achieve and maintain a healthy weight through balanced eating and regular exercise. Weight management is essential for preventing chronic disease and improving overall well-being.

11. **Keep Moving**: Stay physically active through daily exercise and movement. Regular activity boosts circulation and maintains muscle and bone health, enhancing longevity.

12. **Drink in Moderation**: Consume alcohol in moderation to protect your liver and overall health. Reduced alcohol consumption promotes longevity.

13. **Get Spiritual**: Cultivate a sense of spirituality or purpose that gives your life meaning and direction. Spiritual practices can reduce stress and enhance mental and emotional well-being.

14. **Forgive**: Let go of grudges and practice forgiveness to reduce emotional stress and foster healthier relationships. Forgiveness enables mental and emotional longevity.

15. **Use Safety Gear**: Prioritize safety in your activities by wearing appropriate gear (seat belts, ski helmet) and taking precautions. This helps prevent accidents and injuries that can disrupt a long and healthy life.

16. **Make Sleep a Priority**: Prioritize quality sleep by maintaining a consistent sleep schedule and creating a restful environment. Good sleep habits are fundamental for physical and mental rejuvenation.
17. **Manage Stress**: Implement stress-reduction techniques such as meditation, deep breathing, or mindfulness. Effective stress management is crucial for both mental and physical health and longevity.
18. **Keep a Sense of Purpose**: Cultivate a clear sense of purpose in life, whether through work, hobbies, or community involvement. A sense of purpose enhances motivation and resilience, contributing to a fulfilling and longer life.

Appendix B
Abundance360

APPENDIX B

The world is changing faster than ever. Here's how you keep up, survive, and thrive.

You've just finished reading about the incredible breakthroughs in longevity, a future where living healthier, longer lives is within reach. But are you prepared for the other exponential changes coming your way?

The next decade will bring more technological advancements than the last 100 years combined. Artificial intelligence, robotics, virtual reality, quantum computing, and biotechnology are converging to disrupt and reinvent every industry and reshape every aspect of our lives.

Peter converses via Starlink with Elon Musk during the Abundance Summit.

How do you, as a leader, entrepreneur, or investor, not only survive but also thrive?

How do you surf this tsunami of change rather than get crushed by it?

The answer lies in accessing the right *Knowledge* and the right *Community*.

Do you want to gain *Knowledge* about the breakthroughs expected over the next two to three years? This Knowledge comes from an incredible faculty curated by Peter Diamandis at his private leadership summit held each March, the *Abundance Summit*.

Every year, Peter gathers faculty who are industry disruptors and changemakers. Picture yourself learning from visionaries and having conversations with the leaders who have been our faculty over the past years, including Cathie Wood, Elon Musk, Palmer Luckey, Sam Altman, Marc Benioff, Tony Robbins, David Sinclair, PhD, Eric Schmidt, Ray Kurzweil, Andrew Yang, Emad Mostaque, will.i.am, Sal Khan, Arianna Huffington, Michael Saylor, Salim Ismail, Andrew Ng, and Martine Rothblatt (just to name a handful in recent years).

Even more important than Knowledge is *Community*—one that understands your challenges and inspires you to pursue your Massive Transformative Purpose (MTP) and Moonshot(s).

Community is core to Abundance360. Our Members are hand-selected and carefully cultivated—fellow entrepreneurs, investors, business owners, and CEOs running businesses valued from $10M to $10B.

Ray Kurzweil in conversation with Peter Diamandis at the annual Abundance Summit.

Peter supports Abundance members in discovering their Massive Transformative Purpose. He guides Members in shaping their Mindset, helping them create an Abundance, Exponential, Moonshot, Longevity Mindset!

Having the right Knowledge and Community can be the difference between thriving in your business or getting disrupted and crushed by the tsunami of change.

Abundance360 is Singularity University's highest-level leadership program, a year-round community led by Peter Diamandis. A360 provides the insights and support network you need to understand how these technologies will transform our world and how you can use them to grow your business and impact.

> *"I bring my top 20 leaders to Abundance360 and recommend that you do as well; it's about success and that's what we're about."*
> **—Scott Struthers, Founder, William Ray Valentine, LLC**

The year-round Abundance360 membership includes:

- An annual 4 ½ day summit
- Hands-on quarterly Workshops (on longevity and AI)
- Regular Masterminds and forum groups
- One-on-one member matching
- A vibrant, close-knit Community with an uncompromising Mission

APPENDIX B

Admission to Abundance360 is only through invitation or application. If you're interested in joining our Community, visit Abundance360.com to learn more.

"The A360 community has allowed us to achieve things that no other community could have, as fast as we have."
—Tim Nelson, MD, CEO & Founder, HeartWorks

"The A360 mindset has changed the trajectory of my life and my employees, teammates' and family's lives drastically."
—Howard Fineman, CEO, Fineman Global Investment

"Being a member of this community has definitely provided me with an unfair advantage over those who don't see the future like Peter does."
—Jennifer Borislow, President & Founder, Borislow Insurance

Appendix C
Join Peter for His Annual Longevity Platinum Trip – "A Virtual Blue Zone"

APPENDIX C

Interested in the latest longevity breakthroughs described in this book? How AI and epigenetic reprogramming are reversing aging, or how to regrow human organs? Perhaps you're interested in accessing the top longevity-related investment opportunities? Or learning where to find novel treatments for a loved one?

If yes, consider joining Peter Diamandis on his annual Abundance Longevity Platinum Trip. There are three reasons why our members join year after year:

#1. Investment Opportunities: You're an investor or family office looking for early investment access to breakthrough technologies and startups transforming health and longevity.

#2. Personalization: You or a loved one has a particular medical condition, and you want access to the top scientists, technologies, and startups to help address it.

#3. Mindset and Learning: You're fascinated by and passionate about longevity and want to stay on the cutting edge of the advancements adding healthy decades to our lives. Your goal is to enhance your longevity mindset.

*****

"I run a multi-billion dollar biotech investment portfolio. The Longevity Trip gives me access to the cutting edge in science. I've attended every year for the past 6 years. I wouldn't miss a year."
—Daniel Krizek, Portfolio Manager, Citadel (Surveyor)

Peter's 5-day/5-star Longevity Platinum Trips alternate between the East and West Coasts of the US. On odd years (2025, 2027, 2029, etc.), the trip takes place in the Northeast (Boston, Cambridge, New Hampshire). In even years (2026, 2028, 2030, etc.), the trip takes place on the West Coast in various regions in California.

Each trip takes place early in the fall (Sept/Oct) and is capped at 60 participants. This keeps the group size small and intimate, ensuring all participants have full access to all elements of the experience.

Peter will spend all 5 days with you as your private guide and provocateur throughout this personalized, action-packed program. You'll meet with the *top 50 scientists, startup entrepreneurs, and pioneers* and learn about breakthroughs against a wide range of chronic diseases.

PETER DIAMANDIS: LONGEVITY GUIDEBOOK

THERAPEUTICS & TECHNOLOGIES
* CRISPR & Gene Editing
* Gene Therapy & Epigenetics
* Stem Cells & Cellular Medicines
* Organ & Tissue Regeneration
* Vaccines
* AI-driven Drug Discovery
* Imaging & Advanced Diagnostics
* Diet, Sleep, & Exercise
* Nutraceuticals
* Senolytics

DISEASE FOCUS AREAS
* Cardiovascular Disease
* Neurodegenerative Disease
* Cancer Detection/Prevention
* Autoimmune Disease
* Inflammatory Disease
* Infectious Disease
* Gastrointestinal Disease
* Reproductive/Women's Health
* Organ Regrowth/Transplant
* Skin, Hair, & Beauty

In current and past years, our faculty has included many of the leaders in the healthspan and longevity field: George Church, PhD and David Sinclair, PhD (Harvard Medical School); Matt Walker, PhD (UC Berkeley); Shinya Yamanaka, MD, PhD (Gladstone Institutes and Nobel Prize laureate); Eric Verdin, MD (The Buck Institute); Michael Levin, PhD (Tufts University); Dean Kamen (Advanced Regenerative Manufacturing Institute, ARMI); and the leadership of Fountain Life—just to name a few.

Ultimately, Peter's mission is to transform the way you think about longevity by giving you overwhelming evidence of these accelerating health and medical advancements that will add healthy decades to our lives.

Visit Abundance360.com/longevity or scan the QR code above to learn more, see if you qualify, and set up an interview.